I0616254

Beyond the Sound Board

A Practical Guide to Quality Church Audio for Pastors, Leaders, and Media Teams

Dr. Timothy Brandon Farmer

Copyright © 2025 Timothy Farmer

ISBN: 979-8-218-60570-4

DEDICATION

This book is dedicated to everyone who desires
and strives for excellence in ministry.

CONTENTS

ACKNOWLEDGMENTS

To my parents, R.L. and Mary Farmer, your unwavering love, guidance, and prayers have shaped the person I am today. Your wisdom and example have been the foundation of my faith and determination. To my brother, Ronald Farmer, and my sister-in-love, Sylvia Farmer, thank you for your steadfast encouragement and belief in my vision. Your support has meant the world to me.

To my pastor, Dr. William E. Flippin, Sr., and First Lady, Mrs. Sylvia T. Flippin, of The Greater Piney Grove Baptist Church, your leadership, mentorship, and example of faith-driven service have inspired me greatly. Thank you for encouraging me to pursue my passion for technology and ministry.

To the incredible Farmer Tech Group Team, your hard work and dedication have been instrumental in helping me bring transformative solutions to ministries across the nation. Your commitment to excellence and innovation makes everything we do possible.

To all my colleagues in ministry, thank you for continually encouraging me in my work with church technology. Your support, collaboration, and shared vision have motivated me to push boundaries and strive for excellence in serving congregations worldwide.

Lastly, to the more than 500 congregations, pastors, church leaders, and media teams I've had the privilege to train and support, thank you for trusting me with your ministries. Your passion and perseverance to enhance worship experiences through technology are both humbling and inspiring.

This book is a testament to the power of faith, vision, and purpose. Together, we continue to demonstrate that technology can serve as a powerful tool for worship and connection.

INTRODUCTION

In the vibrant tapestry of a church's worship experience, sound is more than just a technical element; it is a profound ministry. From the resonant echo of a choir's harmonies to the subtle nuances of a pastor's sermon, audio shapes the spiritual atmosphere, creating connections that transcend the physical realm. This book, *Beyond the Sound Board: A Practical Guide to Quality Church Audio for Pastors, Leaders, and Media Teams*, is a call to recognize and elevate the vital role of sound in worship, making it not merely functional but transformative.

Why Sound Matters in Worship: Audio as Ministry, Not Just Mechanics

Sound in worship is not incidental; it is foundational. Imagine the profound silence that follows a heartfelt prayer, the collective uplift of spirits during a stirring hymn, or the life-changing impact of a pastor's well-delivered message. These moments hinge on the quality and intentionality of the sound. In many ways, sound is the unseen force that binds the sensory and the spiritual, guiding congregants through the worship journey.

However, in countless churches, sound remains an afterthought—a background operation rather than a central ministry. Poorly balanced microphones, unclear sermons, or overpowering instruments can detract from the worship experience. This book aims to shift that narrative, emphasizing that sound is not merely about buttons and knobs but about creating an environment where worshippers can fully engage with God.

The Real Goal: Enhancing Worship Through Intentional Sound Practices

The ultimate goal of church sound is to facilitate an encounter with the divine. High-quality audio ensures that every spoken word is heard, every note is felt, and every moment is accessible to all, whether in-person or online. Intentional sound practices are about more than technical expertise; they are about serving the congregation's spiritual needs.

Achieving this requires a collaborative effort among pastors, worship leaders, and media teams. Each has a role to play in designing and delivering a worship experience where sound supports and uplifts the message. This book will guide you through the principles, strategies, and tools needed to make this vision a reality.

Audience and Purpose: A Unified Approach for Pastors, Leaders, and Media Teams

This guide is for those who believe in the power of worship—pastors who want their sermons to resonate, worship leaders who dream of seamless music transitions, and media teams striving for excellence behind the scenes. It is a resource for churches of all sizes, from those with modest setups to those with sophisticated audio systems.

The purpose of this book is twofold:

1. To provide practical tools and insights for improving church audio.

2. To inspire a mindset shift where sound is embraced as a ministry integral to worship.

Through its pages, you will learn to bridge the gap between technical know-how and theological understanding, fostering a culture where sound excellence is part of your church's DNA.

Personal Story: The Transformation of Church Audio

In one small church nestled in the heart of a bustling city, the Sunday service was often marred by audio mishaps. The microphone would cut out during sermons, the choir's voices would be drowned out by the organ, and congregants in the back struggled to hear anything at all. Frustration grew among the congregation and the sound team, creating a divide that seemed insurmountable.

Enter a pastor with a vision—a vision of worship where sound would no longer be a distraction but a conduit for divine connection. With a small team of volunteers and a modest budget, the church embarked on a journey of audio transformation. They invested in training, upgraded key equipment, and, most importantly, cultivated a spirit of collaboration. The results were nothing short of miraculous. Sermons became impactful, music filled the sanctuary with clarity, and the congregation felt more connected than ever.

This story, and countless others like it, is proof that investing in sound is an investment in worship. Whether your church is small or large, the principles shared in this book can lead to similar breakthroughs, turning challenges into opportunities for growth and ministry.

Overview of the Book

This book is divided into five parts, each designed to build on the last, creating a comprehensive guide to church audio excellence.

Part 1: Foundations of Quality Church Audio

We begin by exploring the theology of sound in worship, understanding why it matters and how it shapes the worship experience. You will also learn about the essential components of a church sound system and the importance of collaboration between pastors, musicians, and sound engineers.

Part 2: Preparing for Quality Sound

Preparation is key to success. This section covers the role of rehearsals, planning for seasonal audio needs, and effectively leveraging volunteers. These chapters will equip you with practical strategies to anticipate and address audio challenges before they arise.

Part 3: Delivering Quality Sound in Worship

This part dives into the art and science of mixing sound for worship excellence. You will learn how to troubleshoot common audio issues, adapt live sound for streaming, and create an engaging experience for both in-person and virtual audiences.

Part 4: Sustaining a Strong Audio Ministry

The focus here is on sustaining your audio ministry through wise equipment investments, ongoing training, and strong leadership. It also offers practical ways to celebrate the vital role of sound in ministry success.

Part 5: Advanced Practices and Future Trends in Church Audio

In the final section, we explore advanced techniques for exceptional worship sound, designing audio for modern worship spaces, and future trends in church audio technology. From immersive audio to AI-assisted mixing, you will discover ways to keep your ministry on the cutting edge.

A Vision for the Future

As you embark on this journey through the pages of *Beyond the Sound Board*, envision a future where sound becomes a seamless extension of your church's mission. Picture a worship service where every element—from the whispered prayer to the triumphant anthem—is heard, felt, and embraced by the congregation. Imagine a media team that operates with confidence, a worship leader who sings with assurance, and a pastor who preaches with the knowledge that their words will reach every heart.

This vision is not only possible but achievable with intentionality, collaboration, and the practical guidance offered in this book. Together, we can elevate the role of sound in worship, ensuring that it serves its highest purpose: connecting people to God.

A Call to Action

The time to act is now. Whether your church is starting from scratch or seeking to refine its audio ministry, this book is your guide. Let it inspire you to:

- Value sound as a critical part of worship leadership.

- Invest in the training and tools needed for excellence.

- Foster a culture of collaboration and growth within your media ministry.

As you read, reflect on your current audio practices and consider the changes you can make today. Every step toward quality sound is a step toward enriching the worship experience for your congregation. Welcome to *Beyond the Sound Board*. Let's embark on this transformative journey together.

PART 1
FOUNDATIONS OF QUALITY CHURCH AUDIO

In the bustling life of church ministry, audio is often seen as a technical task relegated to the sound booth. However, sound is much more than knobs and cables; it is an integral part of worship that shapes how congregants experience and connect with God. Part 1 of *Beyond the Sound Board* sets the foundation for understanding why sound matters in worship and how it can be approached as a ministry in its own right.

Through these chapters, we will explore the theological grounding for sound in worship, the technical essentials of a church sound system, and the power of collaboration in creating a unified worship experience. This section provides the building blocks for anyone—pastor, leader, or media team member—to elevate the spiritual and practical aspects of church audio.

Chapter 1: Theology of Sound in Worship

Sound is not merely a vehicle for communication; it is a spiritual instrument. Chapter 1 delves into biblical perspectives on sound, music, and worship, showing how sound has been a sacred element throughout history. From the resounding trumpets that brought down Jericho's walls to the hymns sung in early Christian gatherings, sound has always been a conduit for divine connection. This chapter will highlight how intentional sound practices can enhance worship engagement and deepen the spiritual experience of your congregation.

Chapter 2: Understanding the Church Sound System

Every tool in the sound system plays a critical role in delivering clear and effective audio. Chapter 2 demystifies the technical components of a church sound system—microphones, mixers, speakers, and acoustics—and explains how they work together to support worship. This chapter also offers guidance on tailoring sound setups for different worship styles and spaces, ensuring every environment becomes a place of clarity and connection.

Chapter 3: The Power of Collaboration

Church audio is not a solo endeavor. Chapter 3 emphasizes the importance of collaboration between pastors, sound engineers, musicians, and other worship team members. Through practical tips and relatable examples, this chapter will show how effective communication and teamwork can transform sound from a technical necessity into a shared ministry. Building a unified vision for worship sound fosters respect, trust, and excellence within your team.

Part 1 lays the groundwork for approaching sound as a ministry, highlighting the intersection of theology, technology, and teamwork. By understanding why sound matters, the essentials of sound systems, and the dynamics of collaboration, church leaders and teams can set a strong foundation for achieving quality church audio. As we journey further into this guide, these principles will continue to build, leading to practical applications and advanced insights in the chapters ahead.

CHAPTER 1
THE THEOLOGY OF SOUND IN WORSHIP

Sound has the extraordinary ability to shape how we experience the world, and in the context of worship, it holds even greater significance. From the resonant proclamation of the Word to the stirring chords of a hymn, sound carries the messages and emotions of our faith. This chapter explores the theological foundations of sound and its role in enhancing the worship experience, grounded in biblical principles and the spiritual significance of sound as a conduit to the divine.

Biblical Perspectives on Sound, Music, and Worship

The Bible offers profound insights into the role of sound and music in worship. Sound is introduced in the very act of creation, as God speaks the world into existence. Genesis 1:3 records the first divine sound: "And God said, 'Let there be light,' and there was light." This moment emphasizes that sound is not just a medium of communication but also an instrument of creation and transformation.

Throughout scripture, sound serves as a tool for worship, celebration, and divine intervention. The Psalms, often referred to as the hymn-book of the Bible, repeatedly call for joyful noise to the Lord. Psalm 150:3-5 exclaims: "Praise him with the sounding of the trumpet, praise him with the harp and lyre, praise him with timbrel and dancing, praise him with the strings and pipe, praise him with the clash of cymbals, praise him with resounding cymbals." These verses demonstrate the integral role of sound and music in glorifying God and fostering a communal spirit of worship.

The New Testament also underscores the importance of sound in spiritual life. Jesus' Sermon on the Mount (Matthew 5-7) illustrates the transformative power of the spoken word. His teachings, heard by multitudes, exemplify how sound conveys the profound truths of God's kingdom. Additionally, Acts 2 describes the day of Pentecost, when the Holy Spirit descended upon the disciples with a sound like a mighty rushing wind, symbolizing divine presence and empowerment. Such accounts affirm that sound, both natural and orchestrated, is central to God's interaction with humanity.

Beyond these examples, we see sound used to mark pivotal moments in redemptive history. The sounding of trumpets in Revelation announces the unfolding of God's eternal plan. Each blast serves as a clarion call to worship, repentance, and hope. These moments demonstrate the multilayered significance of sound—a medium that transcends mere function and becomes a vessel of divine revelation.

Sound's Impact on Worship Engagement and Spiritual Experience

The power of sound extends beyond its biblical roots to its profound impact on the worship experience. Sound has the ability to unify a congregation, evoke deep emotions, and create a sacred atmosphere that draws people closer to God. For example, a well-mixed choir anthem can inspire awe, while a pastor's clearly articulated sermon can illuminate hearts and minds.

When sound is well-managed, it eliminates distractions and allows congregants to focus on the message and the moment. Conversely, poor audio quality—whether it's microphone feedback, uneven volume levels, or muffled voices—can detract from the worship experience and hinder spiritual engagement. It is the responsibility of the church's media team to ensure that sound enhances, rather than detracts from, the sacredness of worship.

The spiritual implications of sound are particularly significant in fostering accessibility. Thoughtfully designed audio systems, including assistive listening devices, ensure that all members of the congregation, including those with hearing impairments, can participate fully. In this sense, sound becomes a ministry of inclusion and hospitality, reflecting the biblical mandate to care for one another.

Sound also plays a critical role in engaging younger generations. In a world saturated with multimedia experiences, clear and dynamic audio ensures that worship remains relevant and impactful. By investing in audio excellence, churches demonstrate their commitment to reaching all generations with the timeless message of the Gospel.

Sound as a Reflection of God's Order

In 1 Corinthians 14:33, Paul writes, "For God is not a God of disorder but of peace." This verse provides a theological foundation for excellence in sound management. Just as God's creation reflects order and harmony, so should the soundscapes of our worship services. A well-executed sound plan mirrors the divine attributes of orderliness and beauty, creating an environment that honors God and uplifts the congregation.

Sound teams are vital stewards of this order. Their role goes beyond technical skill; it is a spiritual vocation that requires attentiveness, humility, and a heart for worship. By managing audio with precision and care, they enable the Word and music to resonate clearly and meaningfully, facilitating a deeper connection between the congregation and God.

This stewardship extends to the maintenance of audio equipment and systems. Neglecting these tools can result in disruptions that distract from worship, whereas proactive care reflects a commitment to excellence. Regular training, system checks, and updates ensure that the sound ministry remains effective and aligned with the church's mission.

The Theology of Silence

Silence, often overlooked in discussions of sound, holds a profound place in worship. Scripture frequently highlights moments of silence as opportunities for reflection, reverence, and divine encounter. In 1 Kings 19:12, Elijah hears God's voice not in the wind, earthquake, or fire, but in a "gentle whisper." This account reminds us that silence, like sound, can be a medium for experiencing God.

Incorporating intentional moments of silence into worship allows space for personal reflection and spiritual listening. These pauses, supported by thoughtful sound design, enhance the worship experience by providing balance and contrast to the auditory elements of the service. Silence, when paired with sound, creates a dynamic rhythm that mirrors the ebb and flow of spiritual life.

Moreover, silence in worship fosters communal intimacy. A shared moment of quiet creates a sense of unity, as the congregation collectively listens for God's voice. These moments become sacred pauses, allowing the Spirit to move freely and deeply among God's people.

A Call to Excellence in Worship Sound

The theological foundation of sound in worship calls for a commitment to excellence. This commitment is not about achieving technical perfection but about stewarding sound as a gift from God. By investing in quality equipment, training, and collaboration, churches can ensure that their audio ministry honors God and serves the congregation effectively.

Excellence in sound also requires a holistic approach that integrates theology, technology, and teamwork. Pastors, worship leaders, and sound engineers must work together to create a unified vision for worship sound. This collaboration fosters mutual respect and shared responsibility, ensuring that every aspect of the service—from the spoken Word to the final note of the closing hymn—is amplified to the glory of God.

Furthermore, churches must recognize the evolving nature of audio technology. Embracing advancements such as streaming platforms, digital sound boards, and wireless systems allows the church to expand its reach and enhance its ministry. Staying informed and adaptable ensures that the church remains a beacon of excellence in a rapidly changing world.

The theology of sound in worship is both profound and practical. It challenges us to view audio not merely as a technical necessity but as a spiritual ministry that enhances worship and draws people closer to God. By grounding our approach to sound in biblical principles and a commitment to excellence, we can create worship experiences that are both meaningful and transformative.

As we continue this journey through the pages of *Beyond the Sound Board*, let this theological foundation inspire and guide us. The chapters ahead will delve into the practical aspects of church audio, equipping you with the tools and insights needed to deliver quality sound in every worship setting. Together, let us embrace the sacred art of sound, honoring God and serving His people with every note, word, and moment of silence.

CHAPTER 2
UNDERSTANDING THE CHURCH SOUND SYSTEM

Creating a seamless worship audio experience requires more than just equipment; it requires an in-depth understanding of how all components of a sound system work together. This chapter explores the essential elements of a church sound system, highlights how to tailor setups for diverse worship styles and spaces, and provides practical examples to inspire action.

The Building Blocks of a Church Sound System

A church sound system is an interconnected network of devices designed to amplify and enhance audio quality. Mastering these building blocks can transform the worship experience:

1. Microphones

Microphones are the foundation of any sound system, serving as the first point of contact for capturing audio. Key types include:

- **Dynamic Microphones**: Known for durability and versatility, they are ideal for vocals, instruments, and live settings.

- **Condenser Microphones**: These excel at capturing intricate details, making them perfect for choirs or acoustic instruments.

- **Wireless Microphones**: Offering mobility and reducing cable clutter, these are especially beneficial for preachers and vocalists who move across the stage.

- **Boundary Microphones**: Suitable for capturing audio from large groups, such as choirs or congregational singing.

2. Mixers

Mixers are the hub where all audio inputs converge and are balanced. Types of mixers include:

- **Analog Mixers**: Straightforward and cost-effective, suitable for smaller churches with simpler setups.

- **Digital Mixers**: Featuring advanced tools like remote control, scene recall, and precision adjustments, these are ideal for larger churches or those with diverse sound needs.

- **Powered Mixers**: Combine the mixer and amplifier into one unit, simplifying setups for portable or smaller spaces.

3. Speakers and Monitors

Properly configured speakers ensure that sound reaches every corner of the worship space. Types include:

- **Main Speakers**: Deliver sound to the congregation.

- **Subwoofers**: Add depth and richness to music, particularly in contemporary worship.

- **Monitors**: Allow musicians and speakers to hear themselves clearly, reducing errors and enhancing synchronization.

- **Line Array Speakers**: Designed for large spaces, these provide even sound coverage across wide areas.

4. Amplifiers and Signal Processors

These elements power the sound system and shape the output:

- **Amplifiers**: Provide the necessary power to drive speakers at optimal levels.

- **Signal Processors**: Tailor the sound by managing equalization, compression, and special effects to fit the space and worship style.

- **Digital Signal Processors (DSPs)**: Offer advanced processing capabilities for fine-tuning acoustics and integrating with other systems.

5. Accessories and Connectivity

Cables, stands, and other accessories ensure a smooth workflow:

- Invest in high-quality cables to prevent signal interference.

- Use shock mounts and pop filters to enhance microphone clarity and minimize unwanted noise.

- Employ cable management tools like Velcro ties and cable trays to reduce clutter and improve safety.

Tailoring Sound Setups for Worship Styles and Spaces

Churches come in all shapes and sizes, and each requires a unique approach to sound design. Key considerations include:

Worship Styles

- **Traditional Services**: Focus on enhancing spoken word clarity and acoustic music, with minimal effects.

- **Contemporary Services**: Prioritize dynamic range, including bass-heavy music and vibrant vocals, using subwoofers and digital effects.

- **Blended Services**: Combine elements of both, requiring flexible equipment capable of switching between styles seamlessly.

- **Special Events**: Seasonal productions or guest speakers may demand temporary adjustments to equipment or sound profiles.

Worship Spaces

- **Sanctuaries**: These spaces often feature high ceilings and reverberant acoustics, requiring careful placement of speakers and acoustic treatments to reduce echo.

- **Multipurpose Rooms**: Flat, reflective surfaces in these rooms can create challenging acoustics. Portable sound systems with focused speaker placement work well.

- **Outdoor Venues**: Use weather-resistant equipment and ensure proper coverage to combat sound dispersion challenges.

- **Historic Buildings**: Often present unique challenges due to architectural constraints, necessitating creative solutions like hidden speakers or minimal-impact acoustic treatments.

Practical Examples and Scenarios

Small Churches

- **Challenge**: Limited budgets and outdated equipment.

- **Solution**: Prioritize essential upgrades like a mid-range digital mixer and versatile microphones. Engage volunteers for basic training and maintenance. Utilize portable acoustic treatments to enhance sound without significant cost.

Mid-Size Churches

- **Challenge**: Balancing diverse worship styles and accommodating a growing congregation.

- **Solution**: Implement a digital mixer with scene recall to switch seamlessly between service types. Equip musicians with personal monitors for better stage management. Develop a volunteer mentoring program to build expertise and ensure smooth operations.

Large Churches

- **Challenge**: Managing complex setups, including live streaming and multiple audio zones.

- **Solution**: Invest in line array speakers for even sound distribution and advanced mixers with remote control capabilities. Collaborate with professionals to fine-tune acoustics. Employ automation tools to streamline transitions between different elements of a service.

Tips for Success

1. **Team Training**: Volunteers are the backbone of church audio ministry. Regular workshops and mentorship programs can build confidence and expertise. Encourage cross-training to ensure team members can fill in for each other.

2. **Routine Maintenance**: Periodic checks ensure longevity and reliability of equipment. Create a checklist for inspecting cables, testing microphones, and updating firmware. Schedule professional inspections annually for critical components.

3. **Collaboration with Worship Leaders**: Strong communication between tech teams, pastors, and worship leaders ensures alignment on audio goals. Hold pre-service meetings to address last-minute changes or needs.

4. **Documenting Processes**: Detailed records of setups, settings, and troubleshooting steps simplify operations and help onboard new team members. Maintain an inventory of all equipment, noting serial numbers and warranty information.

5. **Leveraging Technology**: Utilize apps and software to manage tasks like sound board settings, live mixing adjustments, and remote troubleshooting.

The Broader Impact of Quality Sound

When sound systems are optimized, they achieve more than technical excellence. They:

- **Enhance Engagement**: Clear and immersive sound captures attention and fosters participation.

- **Amplify the Message**: Every word and note resonates with the intended impact, driving spiritual connection.

- **Elevate Worship**: Quality sound enriches the worship atmosphere, making it memorable and inspiring.

Investing in a comprehensive understanding of church sound systems is an investment in the spiritual and communal health of the congregation. With the right tools, training, and teamwork, churches can create an environment where the message of faith is not just heard but truly felt. By embracing these principles, the church sound ministry can become a cornerstone of worship excellence, setting the stage for transformative experiences for all who gather.

CHAPTER 3
THE POWER OF COLLABORATION

Collaboration is the cornerstone of any successful worship experience, and the sound team plays a pivotal role in this partnership. The ability to synchronize efforts between pastors, worship leaders, musicians, and sound engineers is what transforms technical expertise into a seamless act of ministry. This chapter explores the importance of collaboration, emphasizing the need for mutual respect, clear communication, and practical strategies to foster unity within worship teams.

Pastor and Sound Engineer: Building a Unified Vision

A cohesive worship experience begins with a unified vision. Pastors and sound engineers must work together to ensure the technical aspects of sound align with the spiritual goals of the service.

Understanding Worship Objectives

Pastors bring the theological foundation and overarching goals for worship, while sound engineers provide the technical expertise to execute those objectives. Regular meetings between these two roles can:

- Clarify expectations for sound quality.

- Address specific needs for sermon delivery, such as microphone preferences or sound effects.

- Plan for special services or events requiring unique audio set-ups.

When pastors and sound engineers are on the same page, the service becomes a more cohesive and impactful experience for the congregation.

Building Trust Through Communication

Trust is built through consistent communication. Sound engineers should feel empowered to offer suggestions and solutions, while pastors should be open to feedback. For instance:

- Establishing a weekly check-in to review upcoming services.

- Providing constructive feedback after each service to improve future outcomes.

- Ensuring sound engineers understand the spiritual tone and flow of the service.

Musicians and Sound Engineers: Mutual Respect and Communication

The relationship between musicians and sound engineers is critical to achieving a balanced and harmonious sound. Both roles must recognize their interdependence and work collaboratively to enhance the worship experience.

Preparing Together

Joint preparation sets the stage for success. Musicians and sound engineers should:

- Attend rehearsals together to fine-tune audio levels and address potential issues.

- Share feedback on how sound adjustments impact the overall mix.

- Establish a shared vocabulary for discussing technical and musical elements.

Rehearsals are an opportunity for sound engineers to learn the dynamics of the worship set and for musicians to understand how their contributions fit into the larger soundscape.

Respecting Each Other's Expertise

Mutual respect is the foundation of any successful collaboration. Musicians bring their artistic vision, while sound engineers ensure that vision translates effectively to the congregation. Respecting these distinct roles creates a culture of trust and teamwork.

Practical Tips for Improving Collaboration

Collaboration doesn't happen by chance. It requires intentional effort and strategic planning. Here are some practical tips to improve collaboration before, during, and after services:

Pre-Service Planning

1. **Hold Planning Meetings:** Include pastors, worship leaders, and sound engineers in regular planning sessions to align on goals and expectations.

2. **Use Detailed Run Sheets:** A comprehensive run sheet ensures everyone understands the flow of the service and their responsibilities.

3. **Anticipate Challenges:** Discuss potential technical issues and develop contingency plans.

During the Service

1. **Utilize Clear Communication Tools:** Headsets or walkie-talkies allow for real-time updates and coordination.

2. **Assign Roles:** Designate specific team members to handle adjustments, troubleshoot issues, and monitor sound levels.

3. **Stay Flexible:** Be prepared to adapt to unexpected changes while maintaining professionalism and focus.

Post-Service Evaluation

1. **Conduct Debrief Sessions:** Gather the team after each service to review what went well and identify areas for improvement.

2. **Celebrate Successes:** Recognize the team's hard work and highlight moments where collaboration excelled.

3. **Document Learnings:** Create a record of insights and adjustments to refine future services.

Case Study: Collaborative Success in Worship Sound

A large suburban church struggled with recurring audio challenges that detracted from the worship experience. Miscommunication between the pastor, worship team, and sound engineers often led to inconsistent sound quality. Recognizing the need for change, the church leadership initiated a comprehensive collaboration strategy.

Steps Implemented:

1. **Leadership Commitment:** The senior pastor emphasized the importance of the sound team and established a culture of respect and appreciation.

2. **Regular Training:** The church organized joint workshops for musicians and sound engineers to enhance technical skills and foster teamwork.

3. **Weekly Check-Ins:** A standing meeting was established to review service plans, address concerns, and ensure alignment.

4. **Feedback Loop:** Post-service debriefs allowed team members to share constructive feedback and celebrate successes.

Results:

Within six months, the church experienced a transformation in its worship sound. The congregation noticed a significant improvement in audio clarity and consistency, and the team reported higher morale and stronger relationships. The collaborative approach became a model for other ministries within the church.

Overcoming Common Barriers to Collaboration

Even with the best intentions, barriers to collaboration can arise. Addressing these challenges head-on ensures a healthier dynamic within the worship team.

Misaligned Expectations

Different expectations can lead to frustration and miscommunication. Regular planning sessions and clear documentation help align goals and clarify roles.

Burnout Among Team Members

Sound engineers often work behind the scenes without recognition, leading to burnout. Churches can combat this by:

- Rotating team members to prevent overwork.

- Acknowledging their contributions publicly.

- Providing opportunities for rest and rejuvenation.

Limited Resources

Budget constraints or outdated equipment can hinder collaboration. Transparent discussions about resource limitations and strategic planning for upgrades can mitigate these challenges.

Practical Applications for Continued Growth

To sustain collaboration over time, churches should consider implementing long-term strategies that encourage team growth and development. For example:

- **Training Opportunities:** Schedule periodic workshops and training sessions to keep the team's skills sharp and introduce new techniques.

- **Team-Building Activities:** Organize events outside of regular worship to strengthen relationships and foster camaraderie.

- **Annual Reviews:** Conduct formal evaluations of the sound team's performance and set goals for improvement.

The Role of Conflict Resolution

Collaboration is not without challenges, and conflicts can arise even among well-intentioned teams. Effective conflict resolution is a critical component of maintaining strong collaboration.

Identifying the Source of Conflict

Conflicts often stem from misunderstandings, misaligned goals, or interpersonal dynamics. Encouraging open dialogue helps identify the root cause and find a path forward.

Mediating Discussions

When conflicts arise, leaders should facilitate discussions that allow all parties to share their perspectives. Setting ground rules for respectful communication ensures productive outcomes.

Learning from Disputes

Every conflict is an opportunity for growth. Documenting lessons learned and implementing preventive measures strengthens the team and reduces the likelihood of future disputes.

Leveraging Technology for Better Collaboration

Technology can bridge communication gaps and enhance collaboration within worship teams. Tools such as:

- **Project Management Apps:** Platforms like Planning Center, Trello, or Asana can streamline planning and task assignments.

- **Real-Time Communication Tools:** Apps like Slack or dedicated team radios can improve responsiveness during services.

- **Shared Calendars:** Ensuring everyone is on the same schedule prevents miscommunication and overlapping responsibilities.

The power of collaboration lies in its ability to transform technical tasks into acts of ministry. By fostering unity among pastors, musicians, and sound engineers, churches can create worship experiences that resonate deeply with the congregation. Through mutual respect, clear communication, and intentional planning, the sound team becomes an integral part of the worship ministry, amplifying not just sound but the impact of the gospel message. In this way, collaboration elevates worship to a shared journey that glorifies God and strengthens the community. Moreover, sustained efforts in training, conflict resolution, and technological integration ensure that this collaboration remains effective and adaptable for years to come.

PART 2
PREPARING FOR QUALITY SOUND

As we transition into Part 2 of *Beyond the Sound Board*, the focus shifts from foundational principles to practical preparation—the necessary groundwork for delivering exceptional audio experiences. This section is not merely about the technical aspects of sound but about cultivating the discipline, foresight, and collaboration required to ensure that worship services run smoothly. The chapters in this part offer actionable strategies to optimize rehearsals, anticipate special audio needs, and engage volunteers effectively. Let's preview the journey ahead.

Chapter 4: The Role of Rehearsals

The rehearsal process is more than a warm-up; it is the cornerstone of cohesive worship and sound. Chapter 4 delves into why rehearsals are indispensable for worship teams and sound engineers alike. From step-by-step guides for sound checks to real-world case studies of how rehearsals have enhanced live audio quality, this chapter underscores the value of preparation. Whether addressing vocal levels, balancing instruments, or troubleshooting technical issues, the rehearsal process is the crucible where potential problems are identified and resolved before they impact the worship experience.

Chapter 5: Anticipating Periodic Audio Needs

Churches often face unique audio challenges during seasonal worship events such as Easter, Christmas, or revival meetings. Chapter 5 provides a roadmap for navigating these challenges effectively. From recognizing and planning for specific needs to budgeting for additional equipment or rentals, this chapter equips leaders to be proactive rather than reactive. Practical tips on how to communicate with the sound team and manage resources will empower both leaders and audio technicians to deliver consistent quality, even during high-pressure events.

Chapter 6: Leveraging Volunteers

Volunteers are the lifeblood of many church audio ministries. However, recruiting, training, and retaining these volunteers can be a challenge. Chapter 6 offers strategies for building a sustainable volunteer program. From onboarding processes to defining clear roles and responsibilities, this chapter emphasizes the importance of creating a supportive environment. Furthermore, it provides insights into how to motivate volunteers by recognizing their contributions and offering opportunities for growth. The goal is to build a team that feels valued and equipped to serve with excellence.

Bridging the Gap

Part 2 serves as the bridge between theory and execution, enabling church leaders, pastors, and media teams to translate foundational knowledge into tangible results. By mastering the practices outlined in these chapters, your audio ministry will be better prepared to meet the demands of modern worship services. Whether through rehearsals, advanced planning, or volunteer development, this section provides the tools needed to prepare for success.

As you move forward, remember that preparation is a form of stewardship. The effort invested in planning, rehearsing, and equipping your team is ultimately an investment in the worship experience. Let's prepare to elevate your church's audio ministry to new heights.

CHAPTER 4
THE ROLE OF REHEARSALS

In the journey toward delivering quality church audio, few elements are as critical yet overlooked as rehearsals. Rehearsals serve as the bridge between intention and execution, enabling worship and sound teams to align their efforts for a seamless worship experience. In this chapter, we will explore why rehearsals are indispensable, outline a step-by-step guide to effective sound checks, and present a case study illustrating how preparation transformed a church's live sound quality.

The Importance of Rehearsals

Rehearsals are not merely an exercise in repetition; they are a ministry within themselves. They provide a sacred space for worship teams and sound engineers to collaborate, experiment, and refine. Through rehearsals, teams can uncover potential issues, fine-tune musical arrangements, and ensure that the technical elements of sound align with the spiritual goals of the service.

1. Establishing Unity Rehearsals foster a sense of unity between the worship and audio teams. When musicians and sound engineers work together in a rehearsal setting, they develop an understanding of each other's needs and challenges. This mutual respect is critical for delivering a worship experience that uplifts and engages the congregation.

Collaboration during rehearsals allows for open dialogue, where team members can share insights and ideas. For example, a vocalist might express difficulty hearing the piano in their monitor, prompting the sound engineer to adjust levels. Similarly, a sound engineer might suggest repositioning a microphone for better clarity. These small interactions build trust and create a cohesive team dynamic.

2. Enhancing Technical Precision Every venue presents unique acoustic challenges. Rehearsals allow sound engineers to experiment with microphone placement, equalization settings, and monitor levels. These adjustments are essential to achieving clarity and balance, especially in spaces with challenging acoustics or multiple uses.

The rehearsal process also provides an opportunity to test the integration of new equipment. For instance, if the church has recently upgraded its mixing console, rehearsals offer the perfect setting to familiarize the team with its functionality and capabilities. This proactive approach ensures that all technical elements are optimized for the service.

3. Mitigating Surprises Worship services are dynamic, often incorporating spoken word, musical elements, multimedia, and congregational participation. Rehearsals provide an opportunity to anticipate and address potential technical glitches. From identifying a faulty cable to calibrating wireless frequencies, preparation minimizes disruptions during the actual service.

Rehearsals also allow teams to simulate real-world scenarios. For example, what happens if a microphone suddenly stops working during a sermon? By practicing contingency plans, teams can respond calmly and efficiently to unforeseen challenges.

4. Building Confidence For musicians, vocalists, and speakers, confidence is key. A well-conducted rehearsal instills confidence by familiarizing participants with the flow of the service and the support they can expect from the audio team. Confidence translates into better performance, as team members feel prepared and supported.

Confidence-building extends to the technical team as well. When sound engineers have the opportunity to practice and refine their craft, they approach the service with a sense of readiness and assurance. This confidence is evident in the quality of the audio delivered to the congregation.

Step-by-Step Guide to Effective Sound Checks

A sound check is the cornerstone of any rehearsal. It ensures that the technical foundation for the service is solid, allowing creativity and spirituality to flourish. Here is a detailed guide to conducting an effective sound check:

1. Preparation

- **Arrive Early:** Sound engineers should arrive at least an hour before the worship team to set up and test equipment. Early arrival ensures that there is ample time to address any unforeseen issues without delaying the rehearsal.

- **Inspect Equipment:** Check all cables, microphones, monitors, and mixing consoles for functionality. Equipment inspection should include testing for loose connections, damaged cables, and proper power supply.

- **Review the Service Plan:** Familiarize yourself with the order of worship, including any special elements that may require unique audio setups. Understanding the flow of the service helps in anticipating audio needs.

2. Setting Levels

- **Start with Silence:** Begin by ensuring that all channels are muted. Gradually unmute and test each input one at a time. This methodical approach reduces the risk of feedback and ensures clarity.

- **Gain Staging:** Adjust the gain levels on each channel to optimize signal strength without distortion. Proper gain staging is critical for achieving a clean and balanced mix.

- **Monitor Mixes:** Work with musicians and vocalists to ensure they can hear themselves and each other clearly through their monitors. A good monitor mix enhances performance and reduces strain on performers.

3. Troubleshooting

- **Identify Issues:** Listen for feedback, distortion, or dropouts and address them immediately. Prompt troubleshooting minimizes rehearsal interruptions and builds trust among team members.

- **Replace Faulty Gear:** Keep spare cables, microphones, and batteries on hand to resolve issues quickly. Having backups readily available ensures that rehearsals run smoothly.

- **Frequency Coordination:** For wireless microphones and in-ear monitors, ensure that frequencies are properly coordinated to avoid interference. Frequency conflicts can disrupt both rehearsals and live services.

4. Run the Setlist

- **Play Through Each Song:** Allow the worship team to rehearse their full setlist while the sound engineer adjusts levels and EQ settings for each element. Running the setlist ensures that all audio elements are balanced and cohesive.

- **Balance the Mix:** Ensure that vocals, instruments, and backing tracks are balanced and complement each other. A balanced mix enhances the overall worship experience for the congregation.

- **Simulate the Service Environment:** Adjust the mix for the expected audience size and room acoustics. Anticipating how the space will sound with a full congregation is essential for optimal audio quality.

5. Feedback and Refinement

- **Solicit Input:** Encourage musicians and vocalists to provide feedback on their monitor mixes and overall sound. Open communication fosters a sense of collaboration and ensures that everyone's needs are met.

- **Collaborate:** Work as a team to make final adjustments. Collaboration creates a sense of shared ownership and accountability.

- **Document Settings:** Save presets on digital consoles or take notes on analog settings to replicate the mix during the service. Documentation ensures consistency and reduces the margin for error.

Case Study: Transforming Live Sound Through Rehearsals

At Grace Community Church, sound issues had become a recurring distraction during worship. Feedback from the congregation revealed that the audio was often too loud, unclear, or inconsistent. The media team decided to implement a structured rehearsal process, focusing on sound checks and collaborative planning.

Initial Steps The team began by scheduling weekly rehearsals that included both the worship team and sound engineers. They introduced a standard checklist for sound checks and allocated time for troubleshooting. The initial focus was on building a culture of preparation and excellence.

Challenges Encountered In the first few weeks, the team faced several challenges, including:

- **Resistance to Change:** Some team members were hesitant to adopt the new schedule. Overcoming resistance required clear communication about the benefits of rehearsals.

- **Equipment Limitations:** Outdated equipment presented technical hurdles. Addressing these limitations involved prioritizing upgrades and creative problem-solving.

- **Skill Gaps:** Volunteers lacked experience with advanced mixing techniques. Skill gaps were addressed through targeted training and mentorship.

Overcoming Obstacles Through persistence and collaboration, the team addressed these challenges:

- **Training Workshops:** The church invested in training sessions to equip volunteers with the necessary skills. Workshops covered topics such as gain staging, EQ settings, and troubleshooting.

- **Equipment Upgrades:** They prioritized upgrading key components of their sound system, starting with microphones and monitors. Strategic investments ensured immediate improvements.

- **Celebrating Wins:** Small improvements were celebrated to build morale and encourage buy-in. Positive reinforcement created a sense of momentum and excitement.

Results After three months of consistent rehearsals, the church experienced a noticeable improvement in sound quality. Congregants reported that the audio was clearer and more engaging, and the worship team expressed greater confidence in their performances. The structured rehearsal process also fostered stronger relationships between team members, creating a sense of shared purpose.

The transformation was not limited to technical improvements. The culture of the media team shifted toward one of collaboration, preparation, and excellence. This cultural shift had a ripple effect, inspiring other ministry teams to adopt similar approaches.

Rehearsals are an essential investment in the ministry of sound. By dedicating time to preparation, churches can elevate their worship experiences, minimize technical disruptions, and foster unity among their teams. The lessons learned and skills developed during rehearsals not only enhance the quality of worship but also glorify God through excellence. Whether your church is large or small, the principles outlined in this chapter can serve as a blueprint for building a culture of preparation and collaboration. The next chapter will explore how to anticipate and meet periodic audio needs, ensuring that your church is ready for every season of worship.

CHAPTER 5
ANTICIPATING PERIODIC AUDIO NEEDS

Churches experience a rhythm of seasons and events that significantly influence audio requirements. From major holidays like Easter and Christmas to unique occasions such as revivals, weddings, and funerals, the need for tailored audio solutions arises frequently. Anticipating and addressing these needs proactively is crucial for ensuring quality worship experiences. This chapter explores how churches can recognize periodic audio needs, effectively communicate them, and budget appropriately for additional equipment or services.

Recognizing Seasonal Worship Events

Seasonal events are some of the most predictable yet demanding moments for a church's audio team. These include Easter celebrations, Advent services, and Christmas pageants, among others. Each event carries distinct requirements:

- **Easter:** Often involves a large choir, additional musicians, or even outdoor services requiring portable sound systems.

- **Christmas:** Features musical productions, spoken-word elements, and multi-space setups (sanctuary, fellowship halls, etc.).

- **Revivals:** May involve guest speakers, dynamic preaching styles, and extended services requiring robust microphone and speaker systems.

Steps to Anticipate Needs

1. **Analyze Past Events:** Review audio setups, challenges, and successes from similar events in previous years.

2. **Plan Early:** Engage worship leaders, pastors, and event co-ordinators months in advance to identify special requirements.

3. **Create a Checklist:** Develop a detailed audio checklist tailored to each type of event, ensuring no critical element is overlooked.

4. **Forecast Seasonal Trends:** Anticipate growth in attendance or changes in event formats that could impact sound requirements.

5. **Consider Environmental Factors:** For outdoor events, account for wind noise, ambient sound, and weatherproofing equipment.

6. **Engage in Community Trends:** Understand local cultural and seasonal factors that might influence attendance and participation, adapting audio solutions accordingly.

Communicating Needs Effectively

Effective communication between church leadership, worship teams, and audio teams is the backbone of successful sound management. Here are key practices for ensuring everyone is on the same page:

1. **Regular Planning Meetings:** Schedule pre-event meetings to discuss audio requirements, timelines, and contingencies.

2. **Shared Documentation:** Use shared documents or church management software to track audio needs, equipment lists, and assignments.

3. **Designate Point People:** Assign a liaison from each team (e.g., worship leader, sound engineer) to streamline communication.

4. **Implement Feedback Loops:** Establish mechanisms for teams to share insights and adjustments during rehearsals and live events.

5. **Encourage Open Dialogue:** Create a safe space for discussing potential issues without placing blame, ensuring smoother collaboration during high-pressure events.

The Role of Pastors and Leaders

Pastors and leaders play an essential role in conveying the importance of sound preparation. When leaders prioritize sound quality, it underscores its significance to the congregation and motivates teams to excel. Specific actions leaders can take include:

- **Incorporating Audio in Budget Meetings:** Highlight audio needs during budget discussions to ensure adequate funding.

- **Acknowledging the Sound Team's Efforts:** Publicly recognize their contributions to successful events.

- **Encouraging Cross-Team Collaboration:** Foster a culture where worship leaders, musicians, and audio teams work cohesively.

- **Setting Visionary Standards:** Share a vision of excellence that inspires teams to view their work as ministry, not just technical support.

Budgeting for Additional Equipment or Rentals

Large-scale or seasonal events often necessitate resources beyond a church's standard equipment inventory. Examples include:

- Renting additional microphones, speakers, or mixers.

- Hiring professional sound engineers for events with complex setups.

- Purchasing specialized gear, such as wireless microphone systems or in-ear monitors.

Creating a Budget Plan

1. **Estimate Costs:** Develop cost projections based on anticipated needs. For example, renting a wireless microphone kit might cost $200 for a weekend.

2. **Allocate Funds:** Include a dedicated line item for audio in the church's annual budget.

3. **Build a Contingency Fund:** Prepare for unexpected expenses, such as emergency equipment repairs.

4. **Leverage Donations and Grants:** Explore opportunities for funding through grants or member donations earmarked for audio ministry.

5. **Track Spending:** Maintain detailed records of audio-related expenses to inform future budgeting decisions.

6. **Review Long-Term Investments:** Consider whether rentals or purchases align better with the church's growing needs, evaluating cost-effectiveness over time.

Examples of Successful Planning

Case Study 1: Christmas Musical Extravaganza A midsize church hosting an annual Christmas program faced recurring challenges with insufficient microphones and inconsistent sound quality. After conducting a needs assessment, the church rented additional wireless microphones, conducted a full rehearsal with the sound team, and allocated funds for an experienced audio technician to manage the event. The result was a seamless performance that left attendees praising the clarity and balance of the audio.

Case Study 2: Outdoor Easter Sunrise Service Recognizing that their existing sound system wouldn't suffice for an outdoor service, a congregation invested in a portable PA system. Early testing revealed areas for improvement, leading to adjustments in speaker placement and microphone selection. On the day of the service, the audio team delivered clear sound to over 500 attendees, ensuring the service's impact wasn't diminished by technical issues.

Case Study 3: Revival Series with Guest Preachers A series of revival services featuring guest preachers and musicians required significant audio preparation. The church coordinated closely with the guest teams to understand their preferences for microphone types and monitor levels. Additional equipment was rented to meet these needs, and the sound team scheduled extended rehearsals. The resulting services were dynamic and engaging, with clear, balanced sound enhancing the powerful messages delivered.

Case Study 4: Youth Worship Night A special worship event aimed at engaging the youth ministry introduced unique challenges, including managing high-energy bands and interactive crowd elements. The church invested in additional subwoofers for enhanced bass and consulted with a professional engineer specializing in youth events. The result was a vibrant, engaging atmosphere that resonated with young attendees while maintaining audio clarity.

Building a Culture of Proactive Planning

Anticipating audio needs isn't just a technical exercise; it reflects a church's commitment to excellence in worship. Cultivating a proactive culture involves:

- **Training Teams:** Equip volunteers and staff with knowledge about seasonal audio demands.

- **Reviewing Events Post-Mortem:** After each major event, hold debriefs to discuss what went well and what could be improved.

- **Investing in Long-Term Solutions:** Gradually build a robust inventory of versatile equipment to reduce reliance on rentals.

- **Encouraging Innovation:** Empower the audio team to experiment with new techniques and tools that enhance worship sound.

- **Documenting Processes:** Maintain detailed records of set-ups, equipment used, and lessons learned for future reference.

- **Fostering Mentorship Programs:** Pair seasoned audio technicians with newer team members to transfer knowledge and inspire growth.

Practical Tips for Success

1. **Prioritize Simplicity:** Avoid overcomplicating setups; focus on delivering consistent, reliable sound.

2. **Test Early and Often:** Conduct thorough sound checks well in advance of events to identify and resolve issues.

3. **Engage the Congregation:** Solicit feedback from attendees to gauge the impact of audio quality on their worship experience.

4. **Foster a Team Mentality:** Recognize that audio success is a collaborative effort involving multiple teams and leaders.

5. **Integrate Technology:** Leverage advanced tools like audio analyzers or digital consoles to streamline operations.

By recognizing seasonal patterns, fostering open communication, and budgeting strategically, churches can anticipate and meet periodic audio needs effectively. These practices ensure that technical challenges don't detract from the worship experience, allowing congregations to focus on the spiritual significance of each event. In the end, proactive planning not only supports high-quality sound but also strengthens the broader ministry mission. A well-prepared audio ministry reflects a church's dedication to creating meaningful, impactful worship experiences for all who attend.

CHAPTER 6
LEVERAGING VOLUNTEERS

Volunteers are the lifeblood of many church ministries, and the audio team is no exception. While paid staff or professional engineers may oversee the overall direction, it is often volunteers who ensure the day-to-day execution of the church's audio needs. This chapter explores practical strategies for recruiting, training, structuring, motivating, and retaining volunteers, emphasizing the importance of creating a ministry that thrives on commitment and collaboration.

Recruiting Church Volunteers for Audio Ministry

Finding the right people to serve in the audio ministry begins with intentional recruitment. Churches often have a wealth of untapped talent within their congregations. The key is to identify individuals who not only possess technical skills but also have a heart for service and worship. Here are some effective ways to recruit volunteers:

1. **Make the Need Visible:** Regularly announce the need for audio volunteers during services, in bulletins, or on social media. Share how vital the audio team is to the worship experience.

2. **Host an Information Session:** Offer an open house or behind-the-scenes tour of the sound booth, demonstrating the equipment and explaining the role of the audio team.

3. **Seek Referrals:** Ask current team members, pastors, and ministry leaders to recommend individuals who might be interested.

4. **Engage Younger Members:** Many youth and young adults are naturally tech-savvy. Consider tapping into their skills through youth groups or college ministries.

5. **Highlight the Ministry's Impact:** Share testimonials or stories of how quality sound has enhanced worship. People are more likely to serve when they understand the significance of the role.

Training Volunteers for Audio Excellence

Once volunteers are recruited, training becomes the cornerstone of a successful ministry. Even the most enthusiastic volunteer needs clear instruction and guidance to operate effectively. Here's how to set up a robust training program:

1. **Start with the Basics:** Begin with an orientation that covers the church's sound system, basic audio terminology, and the role of the audio team in worship.

2. **Provide Hands-On Experience:** Schedule regular training sessions where volunteers can practice setting up microphones, mixing sound, and troubleshooting common issues.

3. **Create a Training Manual:** Develop a comprehensive guide that outlines step-by-step instructions for operating the equipment, as well as protocols for rehearsals and services.

4. **Mentorship Program:** Pair new volunteers with experienced team members for on-the-job training and support.

5. **Encourage Continuous Learning:** Provide access to workshops, online courses, or industry resources to help volunteers improve their skills over time.

6. **Simulate Real Scenarios:** Conduct mock services or rehearsals that mimic real-life challenges, such as equipment

malfunctions or sudden changes in the worship lineup. This prepares volunteers for the unpredictability of live audio management.

7. **Celebrate Small Wins:** Acknowledge milestones in training, such as mastering a new skill or successfully managing a service. Positive reinforcement builds confidence and momentum.

Structuring Volunteer Roles for Success

A well-organized team structure is essential to prevent burnout and ensure that all responsibilities are covered. Consider these strategies:

1. **Define Roles Clearly:** Assign specific tasks to each volunteer, such as operating the Sound Board, setting up microphones, or managing livestream audio.

2. **Develop a Rotation Schedule:** Create a schedule that allows volunteers to serve in shifts, ensuring adequate coverage while giving everyone time to rest and worship.

3. **Incorporate a Leadership Hierarchy:** Designate team leaders or captains who can oversee operations, mentor others, and act as liaisons with church leadership.

4. **Set Expectations:** Clearly communicate the time commitment, responsibilities, and expectations for each role. Volunteers are more likely to stay committed when they understand their duties upfront.

5. **Use Technology for Coordination:** Utilize scheduling apps or communication platforms to manage volunteer availability and share updates efficiently.

6. **Encourage Role Rotation:** Allow volunteers to switch roles periodically. This not only prevents monotony but also helps build a well-rounded team that can adapt to various needs.

7. **Document Procedures:** Maintain an updated record of procedures and best practices. This resource can serve as a guide

for both new and seasoned volunteers, ensuring consistency across services.

Motivating and Retaining Volunteers in Ministry

Keeping volunteers engaged and motivated requires intentional effort. People are more likely to remain committed when they feel valued and connected to the ministry's mission. Here are strategies to foster long-term commitment:

1. **Cultivate a Sense of Community:** Encourage team-building activities such as group meals, outings, or prayer sessions to strengthen relationships within the team.

2. **Recognize Contributions:** Regularly acknowledge and celebrate volunteers' efforts through thank-you notes, public recognition, or appreciation events.

3. **Provide Spiritual Encouragement:** Remind volunteers of the spiritual impact of their work, connecting their service to the broader mission of the church.

4. **Offer Growth Opportunities:** Allow volunteers to take on new challenges or leadership roles as they gain experience.

5. **Listen and Adapt:** Solicit feedback from volunteers about their experiences and make adjustments to improve the ministry's operations.

6. **Organize Feedback Sessions:** Schedule periodic meetings to discuss what is working well and identify areas for improvement. Open dialogue fosters trust and collaboration.

7. **Create a Reward System:** Offer small incentives, such as gift cards or personalized thank-you gifts, to show appreciation for consistent and outstanding service.

8. **Encourage Personal Development:** Support volunteers in pursuing external certifications or training programs that align with their interests and roles within the ministry.

Balancing Volunteers and Professional Staff

In many churches, the audio ministry functions as a hybrid model, blending volunteer efforts with professional or part-time staff. Striking the right balance ensures that all aspects of the ministry are covered while maintaining high-quality audio standards. Consider these approaches:

1. **Leverage Professional Expertise:** Use professional staff to handle complex technical issues or oversee major events, allowing volunteers to focus on day-to-day operations.

2. **Foster Collaboration:** Encourage a culture of mutual respect and partnership between staff and volunteers, emphasizing shared goals.

3. **Divide Responsibilities Strategically:** Assign critical tasks to professionals while delegating supportive roles to volunteers.

4. **Invest in Training for All:** Ensure that both volunteers and staff receive regular training to stay updated on the latest audio technologies and practices.

5. **Bridge Communication Gaps:** Establish regular check-ins or meetings between staff and volunteers to align efforts and address any challenges.

6. **Share Success Stories:** Highlight collaborative achievements in church communications, such as newsletters or announcements, to reinforce the value of teamwork.

Case Study: Building a Thriving Volunteer Audio Team

To illustrate the principles outlined in this chapter, consider the example of Faith Community Church. This mid-sized congregation relied entirely on volunteers to manage its audio needs. Initially, the ministry struggled with inconsistent sound quality and volunteer burnout. By implementing a structured training program, defining roles, and fostering a sense of community, the church transformed its audio ministry. Today, Faith Community Church has a dedicated team of 12 volunteers who rotate responsibilities and consistently deliver excellent audio for worship services.

Their success stems from a combination of effective recruitment, ongoing training, and intentional efforts to keep the team motivated. The church also integrated periodic evaluations, ensuring that both equipment and volunteer performance met the growing demands of their ministry. This holistic approach not only improved the worship experience but also strengthened relationships among team members.

Volunteers as Stewards of Sound Ministry

Volunteers are not just helping hands; they are stewards of a vital ministry that directly impacts the worship experience. By investing in recruitment, training, and retention strategies, churches can cultivate a team that is both skilled and spiritually committed. A thriving volunteer audio ministry not only enhances the technical aspects of worship but also strengthens the church community as a whole. With intentionality and vision, every church can leverage the power of volunteers to elevate their sound ministry to new heights.

As the audio ministry grows, it will serve as a testament to what can be achieved when dedication and collaboration are prioritized. Churches that empower volunteers to take ownership of their roles will find that the ministry becomes more resilient, adaptable, and capable of meeting the needs of a dynamic worship environment.

PART 3
DELIVERING QUALITY SOUND IN WORSHIP

The journey to quality sound in worship reaches its crescendo in this section, as we delve into the real-time delivery of sound excellence. Part 3 of this guide is where the theories, preparations, and collaborations discussed earlier come to life, focusing on practical strategies to elevate the auditory experience for congregants both in-person and online.

When the sanctuary fills with voices, instruments, and sermons, the sound team's work becomes the unseen foundation that supports the entire worship experience. Yet, it is here, in the live setting, that challenges often arise. From balancing sound levels across a diverse musical ensemble to resolving technical glitches mid-service, the success of the worship soundscape depends on skill, preparedness, and adaptability.

This section comprises three critical chapters, each designed to empower sound teams, pastors, and worship leaders with the tools they need to deliver exceptional audio in any worship scenario:

Chapter 7: Mixing for Worship Excellence

Mixing is as much an art as it is a science. This chapter provides a step-by-step guide to achieving balanced and dynamic mixes that cater to both the spoken word and music. Whether the service is traditional, contemporary, or blended, you'll learn how to craft a sound that enhances worship for every listener—from the person in the pew to the one watching from their living room.

Chapter 8: Troubleshooting Audio Challenges

Every sound team faces technical issues, but the key is in how they respond. Chapter 8 equips you with practical techniques to quickly address common problems like feedback, distortion, and microphone malfunctions. It also offers insights into staying calm under pressure and preparing contingency plans to ensure that worship continues uninterrupted.

Chapter 9: Audio for Virtual Worship

The world of worship has expanded beyond the walls of the church. With the rise of live streaming and recorded services, this chapter focuses on the unique challenges and opportunities of delivering quality sound to virtual audiences. From choosing the right equipment to mastering audio processing for online platforms, you'll discover how to create a seamless and engaging worship experience for those tuning in from afar.

Together, these chapters provide a comprehensive guide to mastering the moment of delivery—ensuring that the message and music of worship are communicated with clarity, power, and reverence. Part 3 is not just about managing sound; it's about honoring the sacred act of worship through excellence in every note and word.

CHAPTER 7
MIXING FOR WORSHIP EXCELLENCE

In the world of church audio, mixing is both an art and a science. It requires technical knowledge, a deep understanding of the worship context, and the ability to make real-time decisions that impact the worship experience. Whether you're balancing vocals, instruments, or spoken word, mixing is at the heart of delivering a seamless and spiritually engaging worship experience. This chapter will guide pastors, leaders, and media teams through the essential principles and practices for mixing audio in worship settings.

The Foundation: Balancing Vocals, Instruments, and Spoken Word

A well-mixed worship service ensures that every element—from the pastor's sermon to the choir's harmonies—is heard clearly and complements the overall experience. Here are key considerations:

Vocals

Vocals are often the centerpiece of worship. Whether it's a soloist, a choir, or a praise team, ensuring clarity and presence is crucial.

- **Lead Vocals:** Should be prominent without overpowering the mix. Use EQ to enhance clarity by reducing muddiness in the low-mids and adding brightness in the upper mids.

- **Backing Vocals:** Complement the lead without competing. Pan backing vocals slightly left and right to create a wider soundstage.

- **Microphone Selection:** Use microphones that suit the vocalist's range and tone. Dynamic microphones like the Shure SM58 are great for live environments, while condensers may be better for choirs.

Instruments

Instrumental balance can elevate or detract from worship if not handled carefully.

- **Key Instruments:** Determine which instruments are central to the style of worship (e.g., piano and organ for traditional services, electric guitar and drums for contemporary).

- **Frequency Management:** Use EQ to carve out space for each instrument. For example, cut lower frequencies on the piano to avoid clashing with the bass guitar.

- **Volume Dynamics:** Instruments should not overshadow vocals but still maintain presence. Use compression to manage peaks in volume.

Spoken Word

The spoken word—sermons, prayers, and announcements—demands impeccable clarity.

- **Microphone Placement:** Ensure the pastor's microphone is placed properly, whether it's a lapel, headset, or handheld.

- **Feedback Control:** Use a notch filter to eliminate feedback-prone frequencies.

- **Consistency:** Maintain a steady volume level throughout the sermon, using compression as needed.

Tailoring Sound for In-Person and Streaming Audiences

The dual nature of modern worship—serving in-person and streaming audiences—requires a strategic approach to mixing.

In-Person Worship

Acoustics play a significant role in how sound is perceived in the sanctuary.

- **Room Acoustics:** Address challenges like echo and reverb by positioning speakers effectively and using acoustic panels.

- **Audience Interaction:** Monitor audience responses to gauge whether the mix is translating well.

- **Dynamic Range:** Adjust levels to suit the dynamics of live worship, ensuring soft moments are intimate and loud moments are powerful but not overwhelming.

Streaming Worship

Streaming introduces additional variables, such as compression artifacts and different playback devices.

- **Dedicated Mix:** Use an auxiliary mix on the sound board specifically for streaming. This allows you to tailor levels independently of the in-person mix.

- **Consistency:** Streaming platforms often compress audio. Test your mix on multiple devices to ensure it sounds good on phones, tablets, and TVs.

- **Direct Feeds:** Capture instruments and vocals directly from the sound board to minimize ambient noise.

Managing Audio Levels for Dynamic Worship Styles

Different worship styles demand different approaches to mixing. Here are some tips for traditional, contemporary, and blended services:

Traditional Worship

- **Key Elements:** Focus on clarity for choirs, spoken word, and acoustic instruments.

- **Reverb:** Add a touch of reverb to enhance the natural warmth of hymns and choral music.

- **Moderation:** Keep levels moderate to reflect the reverent tone of the service.

Contemporary Worship

- **Key Elements:** Emphasize the rhythm section (drums and bass) alongside electric guitars and lead vocals.

- **Energy:** Use compression and EQ to create a punchy and engaging mix.

- **Effects:** Experiment with delay and reverb on vocals and instruments to enhance the contemporary sound.

Blended Worship

- **Balance:** Merge elements from both traditional and contemporary styles. For example, ensure the piano and drums complement rather than compete.

- **Transitions:** Smooth transitions between styles with gradual adjustments in levels and effects.

- **Flexibility:** Be prepared to adapt quickly to varying musical arrangements.

Practical Tips for Mixing Success

Great mixing is as much about preparation as it is about execution. These practical tips can help ensure a successful mix:

1. **Pre-Service Preparation:**

   - Conduct a thorough soundcheck with all vocalists, musicians, and speakers.

   - Save presets for recurring worship configurations to save time.

2. **Active Listening:**

   - Listen critically to every element of the mix. Use reference tracks to benchmark quality.

3. **Communication:**

   - Maintain clear communication with the worship leader and musicians during the service.

   - Use hand signals or a dedicated communication system to address issues discreetly.

4. **Continuous Learning:**

   - Attend workshops and training sessions to refine your mixing skills.

   - Stay updated on new mixing techniques and technologies.

Advanced Techniques for Mixing Excellence

Once you've mastered the basics, consider incorporating advanced techniques to elevate your mix:

1. **Parallel Compression:**

   • Blend a compressed version of a track with the original to retain dynamics while adding punch.

2. **Subgroup Mixing:**

   • Group similar elements (e.g., all vocals or all drums) into subgroups for easier control and cohesive sound.

3. **Frequency Slotting:**

   • Use EQ to carve out specific frequency ranges for each element to minimize masking and ensure clarity.

4. **Automation:**

   • Automate level adjustments for sections of the service that require different dynamics.

5. **Creative Effects:**

   • Experiment with effects like chorus, flange, and reverb to add depth and interest to the mix.

Case Study: Transforming Worship Sound

At First Community Church, a struggling audio team faced constant complaints about sound quality. Vocals were drowned out, the bass guitar was too loud, and sermons were hard to hear. Recognizing the need for change, the pastor and media team attended a mixing workshop and implemented the following steps:

•   **Invested in Training:** Volunteers learned the fundamentals of EQ, compression, and gain staging.

•   **Collaborated Closely:** Musicians and sound engineers worked together during rehearsals to perfect the mix.

•   **Improved Equipment:** The church upgraded its sound board to include digital mixing capabilities.

The result? A marked improvement in sound quality, leading to a more immersive worship experience and increased congregation satisfaction.

Practical Challenges and Solutions

Even with preparation, challenges can arise. Here's how to address some common issues:

Feedback

- **Solution:** Identify and reduce problematic frequencies using a graphic EQ. Ensure microphones are not placed directly in front of speakers.

Inconsistent Levels

- **Solution:** Use compression to smooth out dynamic variations. Train operators to monitor levels continuously.

Technical Failures

- **Solution:** Have a backup plan in place, including spare cables, microphones, and a portable PA system.

Building a Strong Audio Team

Success in mixing requires a team effort. Invest in the growth and unity of your audio team:

- **Regular Training:** Schedule monthly sessions to practice and discuss new techniques.

- **Team Dynamics:** Foster a culture of respect and collaboration between musicians and audio engineers.

- **Recognition:** Celebrate milestones and achievements to keep the team motivated.

Mixing for worship excellence is about more than technical proficiency—it's about creating an environment where the congregation can connect deeply with God through sound. By balancing vocals, instruments, and spoken word, tailoring mixes for in-person and streaming audiences, and adapting to dynamic worship styles, church audio teams can elevate the worship experience to new heights. With preparation, communication, and a commitment to continuous improvement, any church can achieve mixing excellence that glorifies God and enriches the worship experience for all.

CHAPTER 8
TROUBLESHOOTING AUDIO CHALLENGES

In the dynamic environment of live worship, audio challenges are inevitable. Whether it's a sudden burst of feedback, a malfunctioning microphone, or an unbalanced mix, these issues can disrupt the flow of the service and distract the congregation. For pastors, leaders, and media teams, learning to troubleshoot audio challenges efficiently is not just a technical skill—it's a ministry imperative. This chapter offers practical guidance to address common audio problems, implement backup plans, and maintain composure under pressure.

Common Issues and Quick Fixes

1. Feedback

Feedback is one of the most common audio problems in church sound. It occurs when a microphone picks up sound from a speaker and amplifies it in a continuous loop.

Quick Fixes:

- **Identify the source:** Move around the sanctuary with a handheld microphone to locate the hotspot.

- **Adjust microphone placement:** Ensure microphones are not directly facing speakers or monitors.

- **Use equalization (EQ):** Reduce frequencies between 2kHz and 4kHz, where feedback often occurs.

- **Monitor volume levels:** Avoid excessively high gain settings on the mixer.

Feedback can also stem from poor room acoustics. Churches with highly reflective surfaces like tile or glass may need additional acoustic treatment to minimize this issue. Installing acoustic panels or curtains can make a significant difference in preventing recurring feedback problems.

2. Distorted Sound

Distortion happens when an audio signal is too strong, causing clipping or an unpleasant, fuzzy sound.

Quick Fixes:

- **Lower input levels:** Check gain staging on the mixer to ensure the signal is not peaking.

- **Inspect cables:** Faulty cables can introduce distortion; replace or repair as needed.

- **Assess equipment:** Ensure microphones, DI boxes, and amplifiers are functioning properly.

Distortion can also be the result of mismatched impedance between devices. Ensuring that all components of your sound system are compatible can help reduce this risk.

3. Microphone Problems

Dead microphones, static, or low signal strength can derail a service.

Quick Fixes:

- **Check power sources:** Replace batteries in wireless microphones.

- **Inspect connections:** Secure loose cables and test for broken connectors.

- **Test channels:** Swap microphones to different channels on the mixer to isolate the issue.

Wireless microphones may also encounter interference from nearby devices. Identifying and avoiding overlapping frequencies can eliminate these issues. Many digital mixers include frequency scanning features that simplify this process.

4. Inconsistent Volume Levels

Unbalanced audio—where some voices or instruments dominate while others fade—can disrupt worship.

Quick Fixes:

- **Perform a sound check:** Adjust levels for each input and save the settings.

- **Use compressors:** Smooth out dynamic variations, especially for vocals.

- **Train operators:** Ensure sound team members understand the basics of balancing the mix.

Additionally, real-time adjustments during the service may be necessary. Sound engineers should monitor the house mix actively and make subtle changes as needed.

5. Ambient Noise

Background noise from HVAC systems, outdoor traffic, or congregation chatter can be distracting.

Quick Fixes:

- **Use noise gates:** Minimize unwanted sound when microphones are not in use.

- **Position microphones strategically:** Keep them close to the sound source to reduce ambient pickup.

- **Communicate with the congregation:** Encourage reverence during prayer or reflective moments.

Ambient noise can also be mitigated through strategic scheduling. Holding rehearsals or sound checks during quieter times of day reduces external distractions.

Preparing Backup Plans for System Failures

Even the best sound systems can fail at the worst possible moment. Proactive preparation is essential to minimize disruptions.

1. Redundancy in Equipment

- **Backup microphones:** Have spare wired and wireless microphones readily available.

- **Duplicate cables:** Stock extra XLR, TRS, and power cables.

- **Spare batteries:** Maintain a supply of charged batteries for wireless gear.

Including backup power supplies, such as uninterruptible power sources (UPS), ensures that a power outage doesn't shut down the entire system.

2. Contingency Plans

- **Alternative audio sources:** Keep a portable speaker or amplifier on standby.

- **Manual notes:** Print out worship lyrics or sermon notes in case slides or recordings fail.

- **Communicate proactively:** Notify the congregation with calm assurance if issues arise.

Developing an emergency plan and practicing it with the team can build confidence and reduce response times during an actual failure.

3. Regular Maintenance

- **Inspect equipment:** Schedule periodic checks for mixers, amplifiers, and speakers.

- **Update software:** Ensure digital consoles and streaming platforms are up to date.

- **Clean connections:** Dust and dirt can degrade sound quality; use contact cleaners on connectors.

Regular firmware updates for digital equipment often include bug fixes and new features that enhance performance. Make it a priority to keep systems current.

Staying Calm Under Pressure

When audio challenges arise, the way the team responds can either exacerbate the problem or bring resolution. Maintaining composure is crucial.

1. Developing a Problem-Solving Mindset

- **Stay calm:** Take a deep breath and assess the situation logically.

- **Prioritize tasks:** Address the most disruptive issues first, such as feedback or a dead microphone.

- **Communicate effectively:** Keep the worship leader and pastor informed without causing panic.

2. Training for High-Stress Situations

- **Simulate emergencies:** Practice handling equipment failures during rehearsals.

- **Create a checklist:** Outline troubleshooting steps for common issues.

- **Empower volunteers:** Train the team to act confidently and independently.

Encourage sound operators to think creatively under pressure. Innovative problem-solving can lead to unexpected solutions.

3. Leaning on the Team

- **Delegate roles:** Assign specific troubleshooting tasks to team members.

- **Support each other:** Encourage a culture of collaboration and mutual respect.

- **Seek feedback:** After resolving a problem, discuss what worked well and areas for improvement.

Building a Troubleshooting Toolkit

Every sound team should have a well-stocked toolkit to address issues efficiently. Here are essential items to include:

- **Multimeter:** For testing electrical connections.

- **Cable tester:** Quickly identify broken or faulty cables.

- **Spare cables and connectors:** Ensure compatibility with your system.

- **Contact cleaner:** Maintain clean and reliable connections.

- **Flashlight or headlamp:** Work in dark or dimly lit environments.

- **Notepad and pen:** Document recurring issues for future reference.

- **Digital tuner:** Ensure instruments are properly tuned.

- **Headphones:** Monitor individual channels without affecting the house mix.

Consider adding a small toolkit for basic repairs, including screwdrivers, pliers, and electrical tape.

Learning from Audio Challenges

Audio challenges, while frustrating, present opportunities for growth and improvement. Reflecting on past experiences can strengthen the team's skills and resilience.

1. Post-Service Evaluations

- **Analyze issues:** Identify what went wrong and why.

- **Celebrate successes:** Acknowledge quick resolutions and team effort.

- **Document lessons learned:** Maintain a log of challenges and solutions for future reference.

2. Continuous Education

- **Attend workshops:** Encourage team members to participate in audio training events.

- **Watch tutorials:** Utilize online resources for troubleshooting and sound design tips.

- **Network with peers:** Learn from other church audio teams through forums and conferences.

Continuous education should also include spiritual formation. Encouraging the team to see their work as ministry fosters a deeper connection to their role.

3. Spiritual Perspective

- **Pray for guidance:** Seek wisdom and patience when challenges arise.

- **Focus on ministry:** Remember that quality sound is a tool to enhance worship, not an end in itself.

- **Trust in God's provision:** View setbacks as opportunities to deepen reliance on Him.

Audio issues can serve as reminders to depend on God's strength rather than human expertise. This perspective can transform challenges into moments of worship.

Troubleshooting audio challenges is an integral part of delivering quality sound in worship. By addressing common issues, preparing for system failures, and maintaining composure under pressure, pastors, leaders, and media teams can ensure that technical difficulties do not detract from the worship experience. With the right mindset, tools, and teamwork, every challenge becomes an opportunity to glorify God through excellence in audio ministry. The key is to approach each issue with patience, preparedness, and a commitment to continuous improvement—turning technical hurdles into stepping stones for a more impactful worship experience.

By investing time in training, developing a comprehensive toolkit, and fostering a team culture that values learning and ministry, churches can rise above technical challenges. Each resolution becomes a testimony of God's faithfulness and the team's dedication to serving Him through sound.

CHAPTER 9
AUDIO FOR VIRTUAL WORSHIP

Adapting Live Audio for Streaming and Recordings

In the digital age, virtual worship has become a vital component of church ministry. Whether due to necessity, such as during a pandemic, or to expand the reach of the church beyond physical walls, the ability to deliver high-quality audio to an online audience is paramount. However, adapting live audio for streaming and recordings is not a simple task; it requires intentionality, technical knowledge, and a commitment to excellence.

One of the most significant challenges in transitioning from live to virtual audio is the difference in acoustics and audience perception. In a physical worship space, the congregation benefits from the natural acoustics of the room, supplemented by the sound system. Online, however, the experience is entirely dependent on the audio signal sent through the streaming platform. This makes proper mixing and signal processing crucial.

To begin, it is essential to create a separate audio mix for the online audience. This is often referred to as an "auxiliary mix" or "broadcast mix." Unlike the main house mix, which is tailored to the in-room experience, the broadcast mix ensures clarity and balance for virtual listeners. Investing in a digital mixer with multiple outputs can simplify this process by allowing for separate control of the online mix.

Additionally, it is important to use high-quality microphones and direct inputs to capture sound at its source. Ambient mics, often used to capture room noise for live worship, are less effective for virtual audiences and can lead to a muddy sound. Instead, prioritize direct microphone feeds for vocalists, instruments, and speakers. For keyboards, electric guitars, and other instruments with line outputs, direct feeds eliminate room interference and improve clarity.

Another critical consideration is latency. Streaming platforms introduce delays that can disrupt synchronization between audio and video. Using a reliable audio interface and ensuring proper synchronization during production minimizes these issues. Testing the stream regularly is crucial to catching latency problems before they affect the live broadcast.

Tools and Techniques to Capture High-Quality Sound Online

The tools and techniques used to capture and deliver audio for virtual worship can significantly impact the quality of the final product. Below are key components and practices to ensure success:

1. **Audio Interfaces**: An audio interface serves as the bridge between your sound system and your computer or streaming device. High-quality interfaces reduce noise and distortion, providing a clean signal for virtual worship.

2. **Digital Audio Workstations (DAWs)**: Using a DAW can enhance the quality of your audio production. Tools like EQ, compression, and reverb can be applied to fine-tune the mix and create a polished sound.

3. **Streaming Software**: Platforms like vMix, OBS Studio, or Wirecast offer robust features for managing audio and video streams. Also, use an encoder like a Black Magic Web Presenter if possible to stream without relying on a computer.

4. **Microphones**: Invest in microphones suited for your specific needs. Condenser microphones are ideal for capturing vocals with clarity, while dynamic microphones are more durable and effective in noisy environments.

5. **Acoustic Treatment**: For pre-recorded segments, consider treating your recording space with acoustic panels or foam to reduce echo and background noise. This is especially important for spoken word recordings like sermons or announcements.

6. **Testing and Monitoring**: Before going live, conduct thorough testing of your audio setup. Monitor the stream on multiple devices to ensure consistency across platforms and devices.

7. **Signal Processing**: Use signal processing tools such as equalization, compression, and noise gates to refine the audio quality. Equalization helps balance frequencies, compression ensures consistent levels, and noise gates minimize background noise.

8. **Backup Systems**: Always have backup systems in place. This includes secondary microphones, interfaces, and even a redundant internet connection to ensure seamless production in case of equipment failure.

Creating an Engaging Experience for Virtual Worshippers

High-quality audio is the foundation of virtual worship, but engagement goes beyond technical excellence. Creating a meaningful and immersive experience for online attendees requires intentional planning and creativity.

1. **Dynamic Mixing**: Adjust the mix in real-time to reflect the flow of the service. For example, bring vocals to the forefront during congregational singing and emphasize the speaker's voice during the sermon. This keeps the audio experience dynamic and engaging.

2. **Use of Music**: Music plays a crucial role in worship, and its impact is magnified in a virtual setting. Select songs that resonate emotionally and spiritually with your audience, and ensure that the mix highlights the beauty of the musical arrangement.

3. **Interactive Elements**: Encourage virtual attendees to partici-pate by singing along, typing responses in the chat, or submit-ting prayer requests. The audio team can enhance this experi-ence by ensuring clarity and volume for any prerecorded or live interactions.

4. **Transitions and Effects**: Smooth transitions between seg-ments, such as moving from worship music to the sermon, can maintain focus and minimize distractions. Adding subtle effects like fade-ins and fade-outs can also enhance the flow.

5. **Audience Feedback**: Regularly seek feedback from your vir-tual congregation to identify areas for improvement. Online polls or post-service surveys can provide valuable insights into their experience.

6. **Tailored Audio Content**: Include content specifically de-signed for virtual audiences. This could be a special message from the pastor, a virtual choir performance, or a Q&A ses-sion.

7. **Multilingual Options**: For diverse congregations, consider providing multilingual audio streams or subtitles to engage a broader audience.

8. **Visual Synchronization**: Ensure that the audio aligns per-fectly with the video. This synchronization enhances the overall experience and maintains viewer engagement.

Overcoming Challenges in Virtual Audio Production

While virtual audio production offers unique opportunities, it also presents challenges that must be addressed proactively.

1. **Bandwidth Limitations**: Poor internet connections can result in dropouts or low-quality streams. To combat this, ensure a stable internet connection with sufficient upload speed. Use a wired connection whenever possible for increased reliability.

2. **Budget Constraints**: Many churches operate on tight budgets, making it difficult to invest in high-end equipment. Prioritize purchases based on impact, starting with essential items like microphones and an audio interface. Seek donations or grants to fund additional upgrades.

3. **Team Training**: Volunteers may lack the technical expertise needed for virtual audio production. Provide training sessions and create easy-to-follow guides to empower your team.

4. **Consistency Across Devices**: Audio quality can vary depending on the device and platform used by the audience. Test your streams on multiple devices, including smartphones, tablets, and desktop computers, to ensure consistency.

5. **Sound Fatigue**: Listening to online services for extended periods can lead to sound fatigue for the audience. Balance levels carefully to avoid overly loud or harsh frequencies. Incorporate moments of silence or soft instrumental music to give listeners a break.

6. **Technological Complexity**: Virtual audio setups can be complex. Simplify operations by standardizing processes and using user-friendly software. Document procedures to ensure consistency.

7. **Licensing and Copyright Issues**: Ensure that all music and audio content used in virtual worship comply with copyright laws. Use licensed materials or public domain resources to avoid legal issues.

Case Study: Transforming Virtual Worship Audio

At Grace Community Church, the transition to virtual worship began during the COVID-19 pandemic. Initially, the audio quality was inconsistent, with complaints about unclear vocals and overpowering background noise. Recognizing the importance of audio in virtual worship, the church invested in new equipment, including a digital mixer and condenser microphones.

The audio team also implemented a broadcast mix tailored specifically for online audiences. Volunteers underwent training on mixing techniques and troubleshooting common issues. As a result, the quality of the virtual worship experience improved dramatically, leading to increased engagement and positive feedback from the congregation.

One notable success was the inclusion of interactive elements, such as live prayer requests and virtual communion. The clear and balanced audio mix ensured that these moments were impactful for participants. Today, Grace Community Church continues to prioritize audio excellence in its virtual worship ministry.

Another significant improvement was the addition of multilingual options, allowing non-English speaking members to participate fully. The church's commitment to inclusivity and excellence has made virtual worship an integral part of its ministry.

Virtual worship is an opportunity to extend the church's reach and impact in ways that were previously unimaginable. By focusing on high-quality audio production, churches can create engaging and meaningful worship experiences for online audiences. Whether adapting live audio, utilizing advanced tools and techniques, or fostering creativity, the commitment to excellence in virtual worship audio reflects the heart of the ministry: connecting people to God in every context.

Investing in the right tools, empowering your team, and continually refining the virtual audio experience ensures that your church can effectively minister to a global audience. As technology evolves, so too must the church's approach to virtual worship, embracing innovation while staying true to its mission of spreading the gospel.

PART 4
SUSTAINING A STRONG AUDIO MINISTRY

As churches strive to maintain excellence in worship and ministry, the role of the audio ministry often becomes a vital, ongoing effort rather than a one-time setup. Part Four of this guide is dedicated to equipping churches to sustain and grow a strong audio ministry over time, ensuring that sound continues to enhance worship and connect congregants to the message and music of the gospel.

This section is built around three foundational chapters:

Chapter 10: Investing in the Right Equipment

Every church has unique audio needs, shaped by its size, worship style, and resources. This chapter provides practical guidance on prioritizing equipment upgrades and purchases, ensuring the best balance between cost and quality. It also highlights the importance of long-term planning and collaboration with vendors and consultants to avoid short-sighted decisions. Churches will learn when it makes sense to invest in new equipment and when renting or maintaining existing gear is the better option.

Chapter 11: Training and Empowering Your Team

Sustainable audio ministries are built on a foundation of well-trained and motivated teams. Chapter 11 offers actionable strategies for establishing ongoing training programs, including workshops, tutorials, and hands-on learning opportunities. It emphasizes the value of empowering volunteers and staff to grow in their roles, creating a culture of excellence and collaboration. By investing in the people behind the sound board, churches can ensure consistent and high-quality audio for every service.

Chapter 12: Leadership and Audio Ministry

Strong leadership is essential to the success of any ministry, and audio is no exception. This chapter outlines practical ways pastors and church leaders can support their audio teams, from providing resources and encouragement to celebrating their contributions. It also explores why leaders themselves should develop a basic understanding of sound principles to better advocate for and guide their teams. Recognizing and celebrating the impact of audio ministry can inspire greater commitment and innovation.

Together, these chapters provide a roadmap for churches to move beyond reactive troubleshooting and toward proactive growth and sustainability in their audio ministries. Whether you are a pastor seeking to better support your team, a sound engineer looking for ways to refine your craft, or a volunteer eager to contribute, Part Four offers the tools and inspiration to keep your audio ministry thriving for years to come.

CHAPTER 10
INVESTING IN THE RIGHT EQUIPMENT

When it comes to delivering quality church audio, the foundation isn't just skilled personnel or well-planned worship services—it's the equipment itself. No matter how talented your sound team is, they can only work as effectively as the tools they have. Investing in the right equipment is essential for creating an audio environment that enhances worship, ensures clarity, and invites participation from congregants both in-person and online. In this chapter, we will explore how to prioritize upgrades, work with vendors and consultants, and make smart decisions about when to rent versus buy.

Understanding Your Church's Needs

Before making any investment, it is crucial to understand your church's unique audio needs. This means evaluating your current setup, the size and acoustics of your worship space, and the specific demands of your worship style. For example, a small congregation in a traditional setting may need a vastly different sound system than a larger church with contemporary services.

Assessing Your Current Setup

Take inventory of your existing equipment. Identify what works well and what doesn't meet your expectations. Common issues such as frequent feedback, low-quality streaming audio, or unreliable wireless microphones can all signal areas needing improvement. Engage your sound team to provide insights based on their experiences during worship services.

Defining Worship Goals

Consider your church's worship vision. Are you aiming to improve vocal clarity for sermons? Enhance the musical experience? Expand your virtual worship capabilities? Aligning your goals with equipment choices ensures that every purchase contributes to your mission. Remember to involve key stakeholders, such as pastors, worship leaders, and media team members, in these discussions.

Setting Priorities for Upgrades

Once you have a clear understanding of your needs, prioritize upgrades. Not every piece of equipment needs to be replaced at once. Focus first on the components that will have the most immediate and noticeable impact.

Microphones: The Front Line of Audio

Microphones are often the most critical element of a sound system. Invest in high-quality wired and wireless microphones tailored to specific uses—whether for vocalists, pastors, or choirs. Directional microphones, for instance, help minimize background noise in live and streaming settings. Consider microphones with built-in noise-canceling technology for improved clarity in noisy environments.

Mixing Consoles: Control at Your Fingertips

An outdated mixing console can limit your sound team's ability to deliver quality audio. Modern digital mixers offer features like scene recall, remote control via tablet, and integrated effects that can significantly enhance the worship experience. Evaluate the console's input/output capacity to ensure it meets your current and future needs.

Speakers and Monitors: Delivering Sound Clearly

Evaluate the placement and quality of your speakers and monitors. Are they appropriately positioned to avoid dead zones and uneven sound distribution? Upgrading to powered speakers or line arrays may help achieve consistent coverage across your sanctuary. Additionally, consider investing in in-ear monitors for musicians and vocalists to improve stage sound and reduce overall volume levels.

Streaming and Recording Equipment

For churches engaging in virtual worship, investing in audio interfaces, quality microphones, and software for live streaming is essential. Poor audio quality online can alienate virtual congregants, even if the in-person experience is excellent. Consider cameras and video switchers that integrate seamlessly with your audio setup for a professional online presence.

Balancing Cost and Quality

Church budgets can be tight, so it's important to balance cost with quality. Resist the temptation to buy the cheapest option, as inferior equipment often leads to more frequent repairs and replacements. Conversely, avoid overspending on features that exceed your practical needs.

Research and Reviews

Leverage online reviews and expert opinions to find equipment that offers the best value for your budget. Seek recommendations from other churches with similar needs. Visit trade shows or attend demonstrations to get a hands-on feel for potential purchases.

The Value of Professional Advice

Consider consulting with audio professionals or companies specializing in church sound systems. They can help design an integrated system tailored to your space and goals, saving time and money in the long run. Engage with professionals who have experience in houses of worship to ensure they understand the unique dynamics of your services.

Building Relationships with Vendors and Consultants

Developing a strong relationship with vendors and consultants can streamline the process of upgrading and maintaining your sound system.

Choosing Reliable Vendors

Select vendors who understand the unique needs of church audio. Look for those with a history of working with houses of worship, as they'll likely have insights into acoustics, worship dynamics, and budget constraints. Establish a partnership with vendors who offer ongoing support and training for your team.

Establishing a Maintenance Plan

Work with vendors to establish a maintenance plan for your equipment. Regular check-ups can prevent issues before they arise and extend the lifespan of your system. Keep a log of maintenance activities to track performance over time.

Negotiating Deals and Discounts

Don't hesitate to negotiate pricing or inquire about discounts for non-profits. Many vendors offer special rates for churches or provide bundled deals for equipment purchases. Explore leasing options if upfront costs are prohibitive.

When to Rent Versus Buy

Sometimes, renting equipment is more practical than purchasing outright. This is especially true for seasonal or one-time events such as Christmas productions, revivals, or outdoor services.

Advantages of Renting

- Access to high-end equipment without a large upfront cost.

- Flexibility to use different gear for specific events.

- No need for long-term storage or maintenance.

When Buying Makes Sense

- For equipment used weekly, like microphones and mixers, purchasing is often more cost-effective.

- If your church regularly hosts large events, owning certain pieces of equipment can save money over time. Consider equipment that can be repurposed for multiple applications.

Planning for Long-Term Success

Investing in audio equipment isn't just about meeting immediate needs; it's about setting your church up for success in the future.

Budgeting for Growth

Create a long-term budget for audio ministry. Set aside funds annually for upgrades, repairs, and training to ensure your system evolves with your congregation's needs. Establish an emergency fund for unexpected repairs or replacements.

Training Your Team

Even the best equipment is only as effective as the people using it. Provide ongoing training for your sound team to maximize the potential of new gear. Host workshops, attend webinars, or bring in experts to teach advanced techniques. Encourage your team to stay updated on industry trends.

Embracing Technology Trends

Stay informed about emerging trends in church audio, such as advanced streaming solutions, immersive sound systems, and integration with other worship technologies like lighting and video. Experiment with innovative solutions like ambient mics for congregational singing or AI-assisted mixing tools.

A Balanced Perspective

While investing in the right equipment is vital, remember that it's only one part of the equation. The heart of church audio lies in the people who operate the system and the worship experiences it supports. By making informed, strategic decisions, your church can create an audio ministry that not only sounds great but also glorifies God and serves the congregation effectively.

By approaching equipment upgrades with intentionality and vision, your church can build a strong foundation for audio excellence, ensuring every voice and every note contributes to a powerful worship experience. Recognize that these investments not only improve the technical quality but also create an environment where the message and music resonate deeply with all who participate.

CHAPTER 11
TRAINING AND EMPOWERING YOUR TEAM

Building a robust and effective sound ministry begins with equipping your team. The church audio ministry is often the unsung hero of worship services, creating an environment where congregants can hear the Word, engage with music, and experience God's presence. However, maintaining a high standard requires intentional investment in training and empowering the individuals who make up your sound team. This chapter explores practical strategies to establish an ongoing training plan, offer hands-on learning opportunities, and foster a culture of growth and excellence.

The Importance of Training

Effective training lays the foundation for a sound ministry that operates smoothly and consistently. While technology and equipment play a critical role, it's the people behind the sound board who bring everything to life. Without proper training, even the most advanced systems can falter under unskilled hands.

Bridging the Knowledge Gap

Many churches face a common challenge: recruiting volunteers who have limited or no technical experience. The key to overcoming this hurdle lies in creating an environment where team members feel confident to learn and grow. Training not only enhances technical competence but also instills a sense of purpose and belonging within the ministry.

Volunteers often come from diverse backgrounds and bring unique skills and perspectives. By acknowledging their potential and providing structured pathways to skill acquisition, churches can transform willing participants into proficient team members. Practical strategies include starting with basic equipment orientation sessions and gradually introducing more advanced topics like equalization (EQ) and dynamic range management.

Creating Consistency

Every worship service is unique, but consistency in audio quality is essential. A well-trained team ensures that sound levels, clarity, and balance remain reliable week after week, fostering trust among the congregation and worship leaders. Establishing operational standards, such as a documented workflow for sound checks and equipment setup, helps ensure uniformity in practices regardless of who is serving.

Establishing an Ongoing Training Plan

A one-time training session is not enough to sustain a thriving audio ministry. Instead, churches should develop a continuous learning plan that adapts to the evolving needs of their team and technology.

Assessing Training Needs

Start by evaluating the current skill levels of your team members. Identify areas where they excel and where additional support is needed. This can be achieved through:

- Observing team members during rehearsals and services.

- Conducting anonymous surveys to gather feedback.

- Holding one-on-one discussions to understand individual goals and challenges.

Setting Clear Objectives

Training should align with the overall goals of your sound ministry. Some objectives might include:

- Improving technical proficiency in operating sound equipment.

- Enhancing collaboration with musicians, worship leaders, and pastors.

- Developing troubleshooting skills for common audio issues.

To set these objectives, consider creating a skills matrix that outlines the core competencies required for each role within the sound ministry. For instance, a mixer operator might need advanced knowledge of signal flow and EQ, while a stagehand might focus on cable management and microphone placement.

Scheduling Regular Sessions

Consistency is key to effective training. Schedule regular workshops or sessions, such as:

- **Monthly training meetings:** Focus on specific topics like microphone techniques, mixing fundamentals, or equipment maintenance.

- **Pre-service reviews:** Use sound checks as opportunities to teach best practices in real-time.

- **Annual retreats:** Dedicate a weekend to intensive training, team building, and vision casting.

Furthermore, integrate cross-training opportunities to help team members understand other roles within the ministry. For example, a mixer operator might spend a session assisting with stage setup to gain a broader perspective.

Hands-On Workshops and Tutorials

Interactive, practical learning experiences are among the most effective ways to train sound teams. Workshops and tutorials allow team members to practice skills in a controlled environment, reducing the pressure of live services.

Equipment Demos

Invite a representative from your audio equipment supplier to conduct a demo. This is a great way for team members to:

- Familiarize themselves with new gear.

- Ask specific questions about features and functions.

- Learn tips for optimizing performance.

Additionally, consider organizing field trips to professional audio facilities, such as recording studios or live event venues, to expose your team to industry best practices.

Scenario-Based Training

Simulate real-life situations to prepare your team for the unexpected. For example:

- **Feedback troubleshooting:** Create intentional feedback loops and guide the team through identifying and resolving the issue.

- **Power outages:** Practice switching to backup systems seamlessly.

- **Microphone failures:** Teach how to quickly replace or troubleshoot faulty microphones during a live service.

Incorporating gamification elements, such as timed challenges or friendly competitions, can make these scenarios more engaging and memorable.

Online Learning Resources

Leverage online tutorials, webinars, and courses to supplement in-person training. Platforms like YouTube, LinkedIn Learning, and specialized church audio training websites offer valuable resources that team members can access at their convenience.

Create a shared library of recommended resources and encourage team members to contribute their findings. This collective knowledge base can evolve into a valuable reference for both current and future team members.

Encouraging Growth and Excellence

Beyond technical skills, a thriving sound ministry requires a culture of growth, collaboration, and excellence. This involves nurturing both the technical and spiritual aspects of the team.

Building Confidence

Many volunteers may feel intimidated by the complexity of sound systems or the pressure of live production. Encourage confidence by:

- Celebrating small wins and progress.

- Pairing experienced team members with newer recruits for mentorship.

- Reinforcing the spiritual impact of their role in enhancing worship.

Confidence-building also includes creating a safe space for mistakes. Use errors as teaching moments and emphasize that growth comes from trial and experience.

Fostering Collaboration

Sound ministry doesn't operate in isolation. Strengthen relationships between your sound team and other ministries by:

- Hosting joint meetings with musicians and worship leaders to align goals.

- Creating open channels of communication for feedback and suggestions.

- Participating in team-building activities that promote unity.

In addition to formal meetings, informal gatherings, such as a team lunch or volunteer appreciation night, can strengthen camaraderie and trust.

Recognizing and Rewarding Effort

Show appreciation for the hard work and dedication of your sound team. Simple gestures like thank-you notes, public recognition during services, or small tokens of appreciation can go a long way in boosting morale.

Consider implementing an awards system, such as "Volunteer of the Month," to highlight exceptional contributions and inspire others to excel.

Practical Tools for Empowerment

Equip your team with the right tools and resources to succeed. This includes both technical equipment and educational materials.

Training Manuals and Checklists

Develop a comprehensive training manual tailored to your church's specific setup. Include:

- Step-by-step instructions for operating equipment.

- Troubleshooting guides for common issues.

- Checklists for pre-service, during-service, and post-service tasks.

Access to Professional Development

Invest in professional development opportunities such as:

- Attending audio engineering conferences or workshops.

- Participating in certification programs for sound technicians.

- Networking with other church audio ministries to exchange ideas and experiences.

Feedback and Evaluation

Regular feedback helps team members identify areas for improvement and celebrate successes. Consider implementing:

- **Peer reviews:** Encourage team members to offer constructive feedback to one another.

- **Performance evaluations:** Provide individual assessments based on specific criteria.

- **Post-service debriefs:** Reflect on what went well and what could be improved.

The Spiritual Dimension of Training

The church audio ministry is more than a technical role; it is a form of worship and service. Remind your team of the spiritual impact they have on the congregation's worship experience.

Integrating Prayer and Devotion

Incorporate prayer and devotion into training sessions. Begin meetings with a prayer for guidance and unity, and reflect on scriptures that emphasize the importance of service and excellence.

Encouraging Personal Growth

Support team members in their personal spiritual journeys. This might include:

- Offering resources for Bible study and prayer.

- Encouraging participation in other church activities.

- Providing opportunities for team members to share their testimonies.

Reinforce the connection between their technical work and its spiritual impact. For instance, highlight how clear audio enhances the congregation's ability to engage with sermons and worship.

Training and empowering your team is an ongoing journey that requires commitment, creativity, and collaboration. By investing in their growth, you not only enhance the technical quality of your church's audio ministry but also cultivate a dedicated group of individuals who view their work as an act of worship. With the right tools, a supportive environment, and a shared vision, your sound ministry can thrive and significantly enrich the worship experience for your entire congregation.

CHAPTER 12
LEADERSHIP AND AUDIO MINISTRY

In many churches, the audio ministry is one of the most underappreciated yet essential components of worship. Despite its critical role in delivering God's Word and facilitating worship experiences, sound teams often work in the background with little recognition or support. For pastors and church leaders, investing time and energy in understanding and supporting the audio ministry is not just a technical necessity but a spiritual responsibility.

Why Leaders Should Invest in Sound

A strong audio ministry is foundational to a thriving church. From clear sermon delivery to an immersive worship experience, sound shapes how congregants engage with the service. Leaders who understand this can significantly enhance their ministry's effectiveness.

Consider this: worshippers connect with the message and music through what they hear. Poor audio quality—feedback, muffled voices, or inconsistent levels—can distract and detract from the spiritual experience. Conversely, excellent sound can inspire, uplift, and create a sacred atmosphere that fosters deeper connection with God.

Leadership investment in audio ministry reflects a commitment to excellence in worship. It shows the congregation that every detail, including sound, matters in glorifying God. Such commitment not only enhances the worship experience but also demonstrates to the sound team that their work is valued.

Practical Ways Pastors and Leaders Can Support Sound Teams

Supporting the audio ministry requires more than an occasional pat on the back. It involves intentional actions that foster growth, collaboration, and a culture of excellence. Here are practical ways leaders can engage:

1. **Understand the Basics** Leaders don't need to become audio engineers, but a foundational understanding of sound systems can make a significant difference. Learning about microphones, mixers, and speakers allows leaders to communicate effectively with the team and make informed decisions about upgrades and investments. Consider attending an introductory workshop on church audio or shadowing a sound team member during setup and service. This hands-on experience builds empathy and equips leaders to provide meaningful support.

2. **Establish Clear Communication Channels** Miscommunication is a common source of frustration for sound teams. Pastors, worship leaders, and sound engineers often have different priorities and speak different "languages." Setting up regular meetings between these groups can help align expectations and goals. For example, before a service or event, hold a brief huddle to review the order of worship, special audio needs, and potential challenges. During this time, encourage open dialogue so everyone feels heard and prepared.

3. **Provide Adequate Resources** Budget constraints are a reality for most churches, but leaders must prioritize sound ministry when allocating resources. Investing in quality equipment, training, and maintenance pays dividends in the long run. Work with the sound team to create a wish list of upgrades and tools. Then, incorporate these needs into the church's annual budget planning. Transparent communication about financial priorities builds trust and demonstrates commitment.

4. **Champion Ongoing Training** A well-trained team is a confident team. Leaders can encourage growth by providing access to workshops, webinars, and online courses. Many orga-

nizations and companies offer church-specific audio training resources tailored to varying skill levels.

Consider hosting an annual "Audio Ministry Retreat" where team members can learn from experts, share experiences, and build camaraderie. These events boost morale and skill development.

5. **Celebrate Wins** Recognizing the sound team's contributions fosters a sense of value and motivation. Publicly acknowledge their efforts during services or staff meetings. Celebrate milestones like a flawlessly executed Easter service or the successful installation of new equipment. Simple gestures— thank-you notes, gift cards, or an appreciation dinner—can go a long way in showing gratitude.

Overcoming Common Challenges

Supporting the audio ministry isn't without its challenges. Leaders often face pushback on budget priorities, limited volunteer availability, and misunderstandings about sound's importance. Here are strategies for navigating these obstacles:

1. **Addressing Budget Concerns** Leaders must be proactive in communicating the value of audio investments to the congregation and other stakeholders. Frame these expenditures as integral to the church's mission and ministry effectiveness. Share testimonials or examples of how improved sound has positively impacted worship. Highlight how better sound quality can enhance the overall experience for both in-person and virtual audiences. Use concrete data when available, such as increased online viewership or positive feedback from the congregation.

2. **Recruiting and Retaining Volunteers** Audio ministry demands technical skills, which can make recruiting challenging. Leaders can alleviate this by highlighting the ministry's spiritual significance and offering clear, manageable roles for volunteers. Pair new recruits with experienced team members for mentorship and provide flexible schedules to accommodate other commitments. Retaining volunteers often hinges

on creating a supportive and rewarding environment. Acknowledge their efforts publicly and privately, ensuring they feel valued and integral to the ministry. Additionally, create a structured onboarding program that provides new team members with the tools and training they need to succeed. This could include written guides, hands-on demonstrations, and shadowing opportunities.

3. **Navigating Conflicts** Misunderstandings between sound teams and other ministry leaders are common but avoidable. Promote a culture of mutual respect by fostering open communication and emphasizing the shared goal of excellent worship experiences. When conflicts arise, address them promptly and constructively. Encourage team members to view challenges as opportunities for collaboration rather than sources of division. For example, if a worship leader's preferences clash with the sound team's technical limitations, work together to find a compromise that prioritizes the worship experience.

Recognizing the Vital Role of Audio

Audio ministry is more than knobs, dials, and wires; it's a spiritual service. Sound teams enable the pastor's message to be heard, worship to be felt, and God's presence to be experienced. By recognizing this, leaders elevate audio ministry from a technical task to a core component of worship.

Imagine the impact when pastors, worship leaders, and sound engineers unite with a shared vision. Services flow seamlessly, the congregation engages deeply, and the entire church grows spiritually. This level of excellence requires intentional leadership and unwavering support.

Moreover, consider the broader implications of excellent audio. It enables the church's message to reach beyond the walls of the sanctuary, touching lives through live streams, recordings, and other digital platforms. In an increasingly digital world, sound quality plays a crucial role in how the church's mission is perceived and received.

Leadership as a Catalyst for Change

Church leaders set the tone for the entire congregation. When leaders prioritize sound ministry, it creates a ripple effect. Volunteers feel empowered, congregants experience better worship, and the church as a whole moves closer to its mission.

Here's a simple framework for leaders to follow:

- **Assess:** Evaluate the current state of your audio ministry. Identify strengths, weaknesses, and opportunities for growth.

- **Plan:** Work with the sound team to create a roadmap for improvements. Set achievable goals with clear timelines.

- **Act:** Implement changes step by step. Celebrate small victories along the way to maintain momentum.

- **Review:** Regularly revisit the plan to ensure progress and adapt to new needs.

Additionally, prioritize the integration of audio ministry with other church functions. This might involve coordinating with the media team to ensure consistency across video and audio, or aligning with the worship team to plan seamless transitions during services. Integration fosters a holistic approach to worship that enhances the overall experience.

The Spiritual Dimension of Sound

At its core, audio ministry is about stewardship. Just as churches care for their physical buildings and financial resources, they must also steward the sounds that fill their sanctuaries. Leaders who embrace this perspective elevate sound from a technical necessity to a spiritual responsibility.

Consider incorporating prayers for the sound team into staff meetings or worship services. These acts of inclusion affirm the team's role as co-laborers in ministry. When the sound team feels spiritually supported, they are more likely to approach their work with a sense of purpose and dedication.

Encourage sound team members to view their role as a form of worship. The technical work they do enables the congregation to hear and respond to God's Word. This perspective can transform routine tasks into acts of devotion.

A Call to Action

Leaders hold the power to transform their church's audio ministry. By investing time, resources, and energy, they can create an environment where sound enhances worship rather than hinders it. This transformation begins with a commitment to understanding, supporting, and celebrating the audio ministry.

As you reflect on your role in leading this critical area, ask yourself: How can I better support my sound team? What steps can I take to elevate the quality of our worship sound? The answers to these questions will not only enhance your church's worship experience but also inspire a deeper connection with God for everyone who enters your sanctuary.

Take the first step today. Engage with your sound team, listen to their needs, and begin building a culture that values and prioritizes quality audio. The rewards—for your team, your congregation, and your ministry—will be immeasurable.

PART FIVE
ADVANCED PRACTICES
AND FUTURE TRENDS IN CHURCH AUDIO

As we step into Part Five of this journey, we venture into advanced practices and future trends that shape the soundscapes of modern worship. By now, you have developed an understanding of the foundational principles, effective preparation, and the importance of collaboration in delivering quality church audio. Part Five builds upon these insights, offering a forward-looking perspective for those who aim to elevate their audio ministry to new heights.

A Bridge to the Future

The chapters in Part Four emphasized the significance of sustaining a strong audio ministry by investing in the right equipment, training and empowering your team, and fostering leadership that champions audio excellence. In this section, we go beyond the essentials and delve into the innovative techniques and strategic planning necessary for exceptional worship sound in contemporary and future settings.

Introducing the Chapters

- **Chapter 13: Advanced Techniques for Exceptional Worship Sound**
  This chapter challenges the limits of traditional audio practices by introducing advanced mixing techniques, creative use of effects, and the strategic application of digital signal pro-

cessing (DSP). You will also explore solutions for handling complex scenarios, such as orchestrating sound for large choirs, orchestras, or multi-campus live streams.

- **Chapter 14: Designing Audio for Modern Worship Spaces**
In this chapter, the focus shifts to acoustics and space optimization. From sanctuaries to gymnasiums, outdoor venues, and beyond, learn how to adapt your audio design to the unique demands of various worship environments. The chapter also addresses collaboration with architects and designers during renovations or new construction projects to ensure optimal sound quality.

- **Chapter 15: The Future of Church Audio Technology**
The final chapter explores emerging trends that promise to redefine church audio. From immersive sound technologies and AI-assisted mixing to advancements in wireless systems, discover how to integrate innovation without losing sight of budgetary and practical constraints. This chapter offers insights into future-proofing your ministry, ensuring it remains relevant and effective in the face of rapid technological advancements.

Preparing for the Journey Ahead

Mastering advanced techniques and understanding future trends are not just about staying current—they're about creating an environment where worship is unhindered by technical limitations. By addressing these topics, we equip pastors, leaders, and media teams to craft an audio experience that reflects the beauty and power of worship itself.

As you engage with these chapters, remember that audio ministry is an evolving field. Embrace the opportunity to learn, experiment, and innovate. Whether you are upgrading your equipment, redesigning your worship space, or exploring the possibilities of new technologies, this section aims to guide you toward a future where quality church audio enhances every aspect of the worship experience.

CHAPTER 13
ADVANCED TECHNIQUES FOR EXCEPTIONAL WORSHIP SOUND

The transition from good to exceptional worship sound requires moving beyond basic skills to advanced techniques that enhance the depth, clarity, and emotional resonance of the worship experience. By exploring innovative methods and tools, sound teams can support a profound spiritual connection for congregants.

Beyond Basics: Layering, Effects, and Dynamics

Advanced mixing transforms worship sound from functional to impactful. A well-layered mix ensures that each element—vocals, instruments, and spoken word—has a defined space in the audio spectrum. Here are key techniques:

1. **Layering Instruments and Vocals:**

   - **Frequency Management:** Use equalization (EQ) to carve out distinct frequency ranges for each element. For example, boost mid frequencies for vocals while reducing them for keyboards to prevent masking.

   - **Panning:** Strategically position instruments within the stereo field to create a sense of depth. Drums and bass often anchor the center, while guitars or synths may be panned slightly left or right.

- **Volume Balancing:** Ensure that the most prominent elements, like lead vocals, sit at the forefront while secondary instruments complement rather than overpower the mix. Adjustments should be made dynamically as the service progresses to maintain a cohesive auditory experience.

2. Adding Effects:

- **Reverb and Delay:** These effects can simulate the acoustics of a large sanctuary or an intimate prayer room. Use sparingly for clarity.

- **Chorus and Modulation:** Apply subtle effects to instruments like keyboards to add richness and complexity.

- **Layered Effects Chains:** For more sophisticated results, combine multiple effects. For example, applying a light reverb followed by a delay can create depth without overwhelming the original sound.

3. Dynamic Control:

- **Compression:** Maintain consistent audio levels, especially for dynamic singers or instrumental solos.

- **Automation:** Use digital mixers to program level changes for specific sections, such as crescendos in worship songs.

- **Sidechaining:** Employ this technique to prioritize vocals by subtly lowering instrumental levels whenever the lead singer's voice is active, ensuring clarity and focus.

Handling Complex Audio Scenarios

Church audio teams often face challenges like large choirs, orchestras, and multi-campus live streams. These require careful planning and execution:

1. **Large Choirs:**

   - **Microphone Placement:** Use a combination of overhead and spot microphones to capture balanced sound.

   - **Grouping:** Create submixes for soprano, alto, tenor, and bass sections to allow fine-tuning.

   - **Ambient Recording:** Incorporate room mics to capture the natural blend of voices for a more authentic and immersive choral experience.

2. **Orchestras:**

   - **Instrument-Specific Techniques:** Use close-miking for strings and brass while employing room mics for natural ambience.

   - **Balancing Acoustics:** Adjust EQ to ensure no section dominates, achieving a cohesive sound.

   - **Mix Prioritization:** Highlight key instruments during solos or prominent sections while keeping the overall mix harmonious.

3. **Multi-Campus Live Streams:**

   - **Latency Management:** Use low-latency networking and audio codecs to ensure synchronization between campuses.

   - **Separate Mixes:** Create tailored mixes for in-person and streamed audiences to address different acoustic needs.

   - **Redundancy Planning:** Implement backup systems, such as alternate internet connections or secondary mixing consoles, to ensure uninterrupted live streams.

Leveraging Advanced Tools

The latest tools in audio technology offer powerful ways to refine sound:

1. Digital Signal Processing (DSP):

- **Noise Gates:** Automatically mute microphones when not in use to eliminate background noise.

- **Equalization Filters:** Address problematic frequencies like feedback or room resonances.

- **Advanced Algorithms:** Utilize modern DSP features such as de-essers to manage harsh sibilance in vocal performances or adaptive EQs that adjust in real-time based on the input signal.

2. Compressors and Limiters:

- Use multiband compressors to manage specific frequency ranges, ensuring clarity without overcompression.

- Apply limiters to prevent distortion during high-volume sections.

- **Parallel Compression:** This technique involves blending a compressed version of the audio signal with the uncompressed signal to maintain dynamic range while achieving power and presence.

3. Digital Audio Workstations (DAWs):

- Record and analyze live services for training or improvement.

- Experiment with mixes offline to perfect them before live application.

- **Integrated Automation:** Use DAW software to pre-program cues, effects, and transitions, streamlining live production workflows.

The Human Element: Creativity and Teamwork

Advanced techniques require not only technical expertise but also a collaborative mindset:

1. **Building Intuition:** Encourage team members to develop a "feel" for worship dynamics. Listen critically to live performances and recordings to identify areas for enhancement.

2. **Communication:** Maintain clear dialogue with pastors, musicians, and worship leaders to align sound with the service's spiritual goals.

3. **Training:** Provide ongoing education on advanced tools and techniques. Workshops, webinars, and mentorship programs can help teams stay at the forefront of church audio.

4. **Cross-Training:** Encourage team members to learn multiple roles within the audio ministry to build a resilient and versatile team.

Preparing for Growth: Scaling Your Sound Ministry

As your ministry grows, so do audio challenges. Address scalability proactively:

1. **Expanding Equipment:**

    *   Invest in additional microphones, in-ear monitors, and digital consoles to accommodate larger setups.

    *   Upgrade speakers and amplifiers to ensure sufficient coverage for bigger congregations.

    *   **Future-Proofing:** Opt for scalable systems that can expand as needs evolve, such as modular digital mixers or networked audio systems.

2. Advanced Streaming:

- Use multi-camera video setups integrated with high-quality audio feeds.

- Explore immersive sound options, such as binaural recordings, for a richer online experience.

- **Audience Engagement:** Implement features like real-time audio adjustments based on user feedback for a more personalized streaming experience.

3. Volunteer Development:

- Train specialized roles, such as monitor engineers or streaming technicians, to distribute responsibilities effectively.

- Foster a culture of excellence by celebrating successes and providing constructive feedback.

- **Mentorship Programs:** Pair experienced team members with newer volunteers to build confidence and skill.

Challenges and Opportunities in Advanced Church Audio

Advanced techniques introduce challenges that, if addressed correctly, can transform audio ministry:

1. Overcoming Complexity:

- Break down complex setups into manageable steps. Use templates and checklists to streamline processes.

- **Task Delegation:** Assign specific roles within the audio team to ensure efficiency and accountability.

2. Balancing Innovation with Tradition:

- Embrace new technologies while respecting the unique culture and style of your church.

- **Congregational Education:** Inform the congregation about new audio practices to build understanding and acceptance.

3. **Budget Constraints:**

- Prioritize upgrades strategically, seeking grants or donations if necessary.

- **Community Partnerships:** Collaborate with other local churches to share resources or co-invest in expensive equipment.

4. **Adaptability:**

- Stay informed about emerging technologies and trends, adapting your approach as needed to maintain relevance.

Case Study: A Journey to Exceptional Sound

Consider the case of a mid-sized church transitioning to advanced techniques. By investing in DSP tools and training their team on dynamic mixing, they eliminated feedback issues and enhanced vocal clarity. The result was not only technical improvement but also a noticeable uplift in congregational engagement.

The church's audio ministry expanded to include streaming technicians and monitor engineers, creating specialized roles that streamlined operations. This allowed the team to focus on continuous improvement, including experimenting with immersive sound formats to elevate the worship experience further.

Exceptional worship sound is a journey of continuous learning and growth. By mastering advanced techniques and fostering a culture of collaboration, church sound teams can create transformative worship experiences that resonate deeply with congregants. The pursuit of excellence in sound is, ultimately, an act of ministry — one that amplifies the message and spirit of worship to touch hearts and lives.

Advanced worship sound is not a destination but an evolving practice. Churches committed to excellence will find that their efforts not only improve technical quality but also deepen the spiritual impact of their services. With dedication and creativity, your audio ministry can become a vital cornerstone of worship that bridges tradition and innovation, guiding congregants into meaningful encounters with the divine.

CHAPTER 14
DESIGNING AUDIO FOR
MODERN WORSHIP SPACES

The worship experience has evolved significantly over the years. From traditional church sanctuaries adorned with stained glass and vaulted ceilings to contemporary worship spaces housed in multipurpose buildings, gymnasiums, or even open-air venues, the diversity of environments presents both opportunities and challenges for achieving optimal audio. This chapter explores the principles, strategies, and practical steps necessary to design and maintain audio systems that deliver consistent, high-quality sound across various modern worship spaces.

Understanding Acoustics: The Foundation of Great Sound

Acoustics—the way sound behaves in a given space—is a critical factor in designing audio systems for worship. Poor acoustics can distort music, render spoken words unintelligible, and detract from the overall worship experience. Here are key considerations for understanding and managing acoustics:

1. **Reverberation Time**: The amount of time it takes for sound to decay in a space. Large sanctuaries with hard surfaces often have long reverberation times, which can muddle speech. Acoustic treatments like sound-absorbing panels can mitigate this issue.

2. **Echo and Flutter**: Parallel surfaces can create distracting echoes. Diffusers or strategically placed acoustic panels help break up these sound reflections.

3. **Bass Traps**: Low frequencies tend to accumulate in corners, creating a muddy mix. Installing bass traps can control these problem frequencies.

4. **Room Shape and Materials**: Curved surfaces or irregular shapes can cause uneven sound dispersion. Incorporating materials that balance absorption and reflection ensures clarity.

Practical Tip: Conduct an acoustic analysis of your space using a professional consultant or acoustic measurement software before designing or upgrading your sound system.

To illustrate these principles, consider a large sanctuary with a high vaulted ceiling. The reverberation caused significant challenges during sermons and musical performances. By adding strategically placed sound-absorbing panels and diffusers, the church was able to improve speech intelligibility and enhance the congregation's engagement.

Optimizing Sound for Diverse Worship Spaces

Modern churches often serve as multipurpose facilities, hosting a variety of events. Each setting requires a tailored approach to audio design:

1. **Traditional Sanctuaries**:

    - **Challenge**: Reverberation caused by high ceilings and reflective surfaces.

    - **Solution**: Install directional speakers to focus sound on the congregation while minimizing reflections.

2. **Multipurpose Gymnasiums**:

- **Challenge**: Large open spaces with minimal acoustic treatment.

- **Solution**: Use portable sound systems with line-array speakers to control sound dispersion.

3. **Outdoor Venues**:

- **Challenge**: Unpredictable environmental factors like wind and ambient noise.

- **Solution**: Employ weatherproof speakers and directional microphones to enhance clarity.

4. **Small Group Rooms**:

- **Challenge**: Overwhelming sound from systems designed for larger spaces.

- **Solution**: Use compact speakers and mixers to create an intimate sound environment.

Practical Tip: Plan audio systems with flexibility in mind, allowing easy adjustments for different setups.

Additionally, consider a church that holds outdoor worship services during the summer months. By employing portable, weatherproof speakers and wireless microphones, they ensure the message and music are clearly delivered, creating a meaningful worship experience in an open-air setting.

Collaborating with Architects and Designers

The design and construction of a worship space should involve close collaboration between architects, designers, and audio engineers. Here's how to ensure the best outcomes:

1. **Early Integration**: Involve audio professionals during the initial planning stages. This avoids costly retrofits and ensures the space is acoustically friendly.

2. **Design Considerations**:

   - Use non-parallel walls to reduce standing waves.

   - Include provisions for cable management and equipment racks.

   - Incorporate aesthetic acoustic treatments that blend with the design.

3. **Lighting and Audio Synergy**: Coordinate lighting and audio plans to avoid interference. For example, avoid placing speakers where they might block lighting fixtures or visuals.

Case Study: A mid-sized church in the Midwest engaged an architect and an audio consultant from the outset of their sanctuary renovation. By integrating acoustic panels into the walls and ceilings, they achieved excellent sound clarity without compromising the aesthetic appeal. The project also included hidden cable management solutions, ensuring a clean and professional look.

To maximize collaboration, churches should create detailed blueprints that outline the placement of all audio equipment, lighting fixtures, and acoustic treatments. Regular meetings between architects, designers, and audio teams ensure that all elements align seamlessly.

Strategies for Consistent Sound Across Multiple Venues

For churches with multiple campuses or varied worship spaces, achieving consistent sound quality is vital for a unified worship experience. Consider the following strategies:

1. **Standardized Equipment**: Use the same brand and model of equipment across venues to simplify training and maintenance.

2. **Digital Networking**: Leverage digital audio networks like Dante to share audio resources between locations seamlessly.

3. **Preset Configurations**: Save sound board presets for different worship styles or venues to ensure consistency.

4. **Regular Calibration**: Conduct periodic sound checks and calibrations using reference tracks to maintain quality.

Practical Tip: Train your audio team to understand the nuances of each venue and adapt settings accordingly.

A large multi-campus church in Texas uses standardized digital mixers and shared audio networks. This allows them to train their volunteers centrally and apply uniform settings across all locations, ensuring a consistent experience for attendees regardless of venue.

Embracing Technology for Modern Sound Design

Emerging technologies offer innovative solutions for designing audio systems in modern worship spaces. Key advancements include:

1. **Beamforming Microphones**: These microphones focus on sound from specific directions, minimizing ambient noise.

2. **Immersive Audio**: Systems like L-ISA create a three-dimensional soundscape, enhancing the worship experience.

3. **Wireless Connectivity**: Advances in wireless technology allow for flexible microphone and speaker placement.

4. **AI-Assisted Mixing**: AI tools can analyze and adjust mixes in real time, reducing the learning curve for volunteers.

Case Study: A large urban church implemented an immersive audio system, creating an enveloping sound experience for its congregation. Feedback from attendees highlighted a deeper sense of connection during worship. Additionally, the system's intuitive interface enabled volunteers to manage complex setups with ease.

Practical Tip: Evaluate emerging technologies based on your church's specific needs and budget. Pilot new tools in one venue before implementing them broadly.

Practical Steps for Implementation

Implementing effective audio design in modern worship spaces requires a structured approach:

1. **Assessment and Planning**:

   - Conduct a needs assessment with input from pastors, musicians, and sound engineers.

   - Create a detailed budget and timeline.

2. **Selecting the Right Team**:

   - Hire experienced audio consultants and contractors.

   - Train an in-house team to maintain and operate the system.

3. **Testing and Tuning**:

   - Perform comprehensive testing before the system's debut.

   - Use reference tracks and congregation feedback to fine-tune settings.

4. **Ongoing Maintenance**:

   - Schedule regular maintenance checks.

   - Update software and firmware to ensure compatibility.

Churches can also create a maintenance log, documenting all updates, repairs, and calibration sessions. This ensures accountability and helps in identifying recurring issues.

Inspiring a Vision for Quality Sound

Ultimately, the goal of designing audio for modern worship spaces is to create an environment where the message and music resonate deeply with the congregation. By understanding acoustics, leveraging technology, and fostering collaboration, churches can enhance their worship experience and ensure that every word and note is heard clearly.

Practical Tip: Cast a vision for quality sound as a ministry priority, emphasizing its impact on worship engagement and spiritual growth.

Consider conducting workshops or presentations to educate church leadership and the congregation about the importance of audio in worship. Sharing testimonies and success stories can inspire further investment and support for the audio ministry.

CHAPTER 15
THE FUTURE OF CHURCH
AUDIO TECHNOLOGY

The landscape of church audio technology is evolving rapidly. With advancements in immersive sound, artificial intelligence, and wireless systems, churches are presented with unparalleled opportunities to enhance worship experiences. This chapter delves into emerging trends, offers strategies for future-proofing audio ministries, and discusses balancing innovation with budgetary and practical needs.

Emerging Trends in Church Audio Technology

1. Immersive Audio Systems

Immersive audio technology is transforming how worshippers experience sound. Unlike traditional stereo or mono setups, immersive systems create a three-dimensional sound field, enveloping the audience. This technology brings:

- **Heightened Engagement**: Sound emanates from all directions, simulating a more natural acoustic environment.

- **Enhanced Clarity**: Each instrument and vocal is placed in a virtual "space," minimizing overlap and maximizing clarity.

- **Practical Applications**: Churches can use immersive sound for dramatic presentations, choir performances, and even congregational singing.

Brands like L-ISA and d&b audiotechnik's Soundscape are leading the charge, making these systems increasingly accessible to houses of worship. Churches utilizing immersive audio have reported greater congregational participation and deeper engagement with worship services, emphasizing how such systems foster spiritual connection.

2. AI-Assisted Mixing

Artificial intelligence is reshaping how audio is managed during worship services. AI-powered tools assist sound engineers by:

- **Automatic Balancing**: AI can adjust levels dynamically based on input sources, ensuring consistent sound.

- **Feedback Suppression**: Algorithms can detect and eliminate feedback in real time.

- **Learning Capabilities**: Some systems "learn" your church's audio profile, improving performance over time.

Software like iZotope and platforms integrated into digital mixers are examples of AI-driven solutions already in use. AI tools reduce the pressure on less-experienced sound technicians, making high-quality sound achievable even in smaller churches with limited expertise.

3. Advancements in Wireless Technology

Wireless audio is no longer just a convenience; it is a necessity. Modern advancements include:

- **Digital Transmission**: Reduced latency and interference compared to older analog systems.

- **Extended Battery Life**: Rechargeable systems with longer runtimes to minimize interruptions.

- **Flexible Connectivity**: Easier integration with multiple devices and platforms.

Future trends point to even more reliable and secure wireless systems, essential for maintaining high-quality sound in complex worship setups. Churches should prioritize upgrading to digital wireless systems, especially in urban areas where frequency interference is more common.

4. Integration of Streaming Technologies

Streaming has become an integral part of church audio ministries. Enhanced tools for live audio capture include:

- **Dedicated Streaming Mixes**: Separate mixes optimized for online audiences ensure clear and balanced sound.

- **High-Resolution Audio**: Streaming platforms now support superior audio quality, providing an immersive experience for remote worshippers.

- **Synchronization Tools**: Technologies that ensure audio and video alignment improve the overall quality of virtual services.

Future-Proofing Your Church's Audio Ministry

Planning for the future doesn't just mean buying the latest equipment. It requires a strategic approach to ensure longevity and adaptability.

1. Assess Current Needs and Limitations

Start by evaluating your church's existing audio setup:

- **Identify Weak Points**: Is there consistent feedback? Are certain areas of the sanctuary underserved acoustically?

- **Gauge Usage Patterns**: How often is the system used for non-worship events? What about streaming?

- **Survey the Congregation**: Gather feedback from attendees to understand how they perceive the sound quality.

2. Set Realistic Goals

Develop a roadmap based on your church's vision and budget:

- **Short-Term**: Address immediate issues like mic placement or speaker alignment.

- **Medium-Term**: Upgrade components like mixers or amplifiers.

- **Long-Term**: Transition to cutting-edge technologies, such as immersive audio or AI-assisted mixing.

3. Invest in Scalable Technology

Scalability is crucial. Choose systems that:

- **Accommodate Growth**: Ensure the equipment can handle larger congregations or additional instruments.

- **Integrate Easily**: Opt for solutions compatible with your current setup.

- **Adapt to New Needs**: Look for modular systems that allow incremental upgrades.

Scalable technology also ensures flexibility, allowing your audio set-up to adapt to changes in worship style or venue requirements without requiring a complete overhaul.

Balancing Innovation with Budget Constraints

One of the biggest challenges churches face is incorporating advanced technology without overextending financially. Here are strategies to maintain a balance:

1. Prioritize Key Investments

Focus on the upgrades that will have the most significant impact:

- **Mixers**: A digital mixer with scene recall and remote control capabilities can revolutionize sound management.

- **Speakers**: High-quality speakers tailored to your space ensure clear sound projection.

- **Microphones**: Reliable, versatile microphones for pastors, singers, and instrumentalists are essential.

2. Leverage Grants and Donations

Many organizations and congregants are willing to support audio ministry enhancements:

- **Apply for Grants**: Look for programs that fund technological advancements in houses of worship.

- **Launch Fundraising Campaigns**: Clearly communicate the benefits of the upgrades to inspire contributions.

- **Encourage Donations**: Accept in-kind donations of gently used equipment from local businesses or members.

3. Partner with Vendors

Establish relationships with audio vendors and consultants who understand the unique needs of churches. Negotiate:

- **Bulk Discounts**: For purchasing multiple items like microphones or speakers.

- **Rental Options**: For seasonal events requiring additional equipment.

- **Maintenance Packages**: To ensure the longevity of your investments.

4. Consider Shared Resources

Collaborate with nearby churches to share audio equipment and expertise. This approach can reduce costs and foster a sense of community.

Practical Steps for Implementation

To bring future-oriented audio solutions to life, consider these practical steps:

1. Engage Your Team

Involve sound engineers, musicians, and leaders in the decision-making process. Their insights can:

- **Identify Pain Points**: They're on the frontlines of audio challenges.

- **Foster Buy-In**: Participation ensures smoother adoption of new systems.

2. Test New Technologies

Before committing to a purchase:

- **Request Demos**: Many vendors offer trial periods or demonstrations.

- **Evaluate Performance**: Test equipment during a live service to gauge real-world effectiveness.

- **Solicit Feedback**: Ask the congregation and team members about noticeable differences.

3. Train Your Team

Equip your volunteers and staff with the knowledge they need to excel:

- **Workshops and Seminars**: Host sessions led by industry experts.

- **Online Tutorials**: Share video resources for ongoing learning.

- **Hands-On Training**: Practice sessions with new equipment to build confidence.

4. Document Processes

Create manuals and checklists for:

- **Equipment Operation**: Step-by-step guides for common tasks.

- **Troubleshooting**: Quick solutions for recurring issues.

- **Maintenance**: Schedules for cleaning and servicing equipment.

Documentation not only ensures consistency but also makes onboarding new volunteers easier.

Building a Vision for Long-Term Success

A successful audio ministry requires more than technical expertise; it demands a shared vision of excellence:

- **Cultivate Leadership**: Encourage sound team leaders to mentor and inspire others.

- **Foster Collaboration**: Maintain open communication between the audio team, worship leaders, and pastors.

- **Celebrate Milestones**: Recognize and appreciate the contributions of your audio team regularly.

The future of church audio is bright and full of possibilities. By staying informed about emerging trends, making strategic investments, and involving your team in the journey, your church can create an audio ministry that resonates both spiritually and technologically.

Remember, the ultimate goal is not just high-quality sound but fostering an environment where worship can flourish. As you embrace these advancements, you're not just investing in technology but in the spiritual lives of your congregation.

CONCLUSION
BEYOND THE SOUND BOARD

As the final notes of a hymn reverberate through a sanctuary or a pastor's message reaches the ears of congregants with clarity and depth, the importance of sound in worship becomes undeniable. This book, "Beyond the Sound Board: A Practical Guide to Quality Church Audio for Pastors, Leaders, and Media Teams," has sought to illuminate the pivotal role audio plays in worship and provide a roadmap for churches to harness its transformative power. Now, as we draw to a close, let us reflect on the journey we have taken and the vision we can carry forward.

A Call to Action: Valuing Sound as Worship Leadership

Sound is not merely a technical aspect of worship; it is a ministry in its own right. When audio quality falters, the ability of a congregation to fully engage with worship diminishes. Conversely, when sound is intentional and seamless, it fosters a profound connection between the message, music, and worshippers. As leaders—whether pastors, worship leaders, or media team members—it is essential to recognize that sound is a critical component of worship leadership.

Churches must shift their mindset to see audio not as a background function but as a frontline ministry. This requires investment, education, and a commitment to excellence. The call is clear: to elevate sound as a vital part of worship leadership and ensure that it enhances the spiritual experience for all.

Practical Steps for Immediate Implementation

Change often begins with small, deliberate steps. Here are practical actions churches can take to begin transforming their audio ministry:

1. **Assess Current Sound Systems:** Conduct a thorough evaluation of your church's sound equipment and identify areas for improvement. Engage professional consultants if necessary.

2. **Build Collaboration:** Foster open communication between pastors, musicians, and sound engineers. Schedule regular meetings to align on worship goals and troubleshoot challenges.

3. **Invest in Training:** Offer workshops and training sessions for media teams. Equip volunteers and staff with the knowledge they need to operate sound systems effectively.

4. **Prioritize Sound Checks:** Emphasize the importance of rehearsals and sound checks. Develop a consistent process to ensure that audio quality is tested and optimized before every service.

5. **Embrace Technology:** Leverage digital tools and resources to improve sound quality for both in-person and virtual worship. Research and implement software or hardware solutions that suit your church's needs.

6. **Celebrate Progress:** Recognize and celebrate the efforts of your sound ministry team. Acknowledge their contributions during worship and encourage them to keep striving for excellence.

Vision for the Future: A Church Culture That Prioritizes Sound

Imagine a future where every church, regardless of size or budget, prioritizes audio as a key element of worship. In this vision, pastors understand the nuances of sound systems, sound engineers view their role as ministry, and congregants experience worship without distraction. This is not an unattainable dream but a realistic goal that requires collective effort and shared commitment.

To achieve this vision, churches must:

- **Adopt a Long-Term Perspective:** View sound ministry as an evolving discipline that requires ongoing investment and innovation.

- **Cultivate a Culture of Excellence:** Encourage a mindset of continuous learning and improvement within the audio ministry.

- **Foster Intergenerational Involvement:** Involve individuals of all ages in sound ministry, blending youthful enthusiasm with seasoned expertise.

- **Expand Accessibility:** Strive to make worship accessible to all, including those with hearing impairments, through thoughtful audio solutions.

Sound as a Ministry: Deepening Spiritual Impact

Sound ministry extends beyond technical expertise. It plays a spiritual role in creating an environment where worshippers can encounter God. Every microphone adjustment, speaker placement, or audio mix contributes to the sacred atmosphere of worship. It is, in essence, a form of stewardship—carefully managing resources to magnify the worship experience.

Consider the impact of sound on a worship service: the resonant bassline of a hymn that stirs hearts, the crisp articulation of a sermon that brings clarity to complex theological truths, or the seamless transitions between spoken word and music that maintain the flow of worship. Each of these elements relies on intentional sound practices that prioritize the worshipper's experience.

Challenges as Opportunities

The path to achieving quality church audio is not without obstacles. Technical failures, budget constraints, and volunteer turnover are common challenges. However, these challenges present opportunities for growth, innovation, and collaboration.

For example, a microphone failure during a service might prompt a church to invest in higher-quality equipment or train volunteers in troubleshooting techniques. Budget limitations can inspire creative solutions, such as forming partnerships with local businesses or leveraging community resources. Volunteer turnover can become an opportunity to recruit new members, bringing fresh perspectives and energy to the sound ministry.

Gratitude for the Sound Ministry Teams

This book would be incomplete without acknowledging the unsung heroes of worship services—the sound ministry teams. These dedicated individuals often work behind the scenes, ensuring that every voice, instrument, and melody reaches the congregation as intended. Their commitment to excellence and willingness to serve are invaluable to the life of the church.

Pastors and leaders, let us not take these teams for granted. Express gratitude, offer support, and empower them to succeed. A thriving sound ministry is a testament to the collective efforts of a church community.

Building a Legacy of Excellence

The pursuit of excellence in sound ministry has implications that extend far beyond the walls of a sanctuary. It establishes a legacy of intentionality and care that future generations can build upon. By investing in quality audio practices today, churches lay the foundation for vibrant worship experiences that inspire and transform lives for years to come.

This legacy also involves passing down knowledge and skills to the next generation. Churches can achieve this by documenting sound practices, mentoring young volunteers, and creating opportunities for intergenerational collaboration. In doing so, they ensure that the principles of quality sound ministry endure.

Fostering Community Through Sound

At its heart, worship is about creating connections—between individuals and God, between members of the congregation, and between generations. Quality sound enhances these connections by ensuring that every word spoken and note played is accessible to all. It fosters an inclusive environment where everyone, regardless of age or ability, can participate fully in worship.

Sound ministry also has the power to build bridges beyond the church walls. Streaming services, podcasts, and other digital platforms extend the reach of worship to those unable to attend in person. They offer a lifeline to shut-ins, travelers, and seekers who might not otherwise engage with the church community. In this way, sound ministry becomes a tool for evangelism and outreach, amplifying the church's message to a global audience.

Embracing Change and Innovation

The world of audio technology is ever-evolving, offering new tools and techniques to enhance worship experiences. Churches must embrace this change with a spirit of curiosity and openness. Staying current with audio trends and innovations not only improves the quality of worship but also demonstrates a commitment to excellence.

However, embracing change does not mean abandoning tradition. The goal is to strike a balance between preserving the sacred elements of worship and incorporating modern advancements. By doing so, churches can create a worship experience that resonates with both long-time members and newcomers.

Cultivating a Spirit of Service

Sound ministry is ultimately about service—to God, to the church, and to the community. It requires humility, dedication, and a willingness to put the needs of others first. Those who serve in this ministry often do so without recognition or fanfare, yet their impact is profound.

Encouraging a spirit of service within the sound ministry team can lead to greater unity and purpose. When team members view their work as an act of worship, they are more likely to approach it with care and intentionality. This perspective transforms technical tasks into acts of devotion, infusing the ministry with deeper meaning.

A Unified Vision

The journey "Beyond the Sound Board" calls us to unity. When pastors, leaders, and media teams come together with a shared vision, the result is transformative. Worship becomes more than an event; it becomes an encounter with the sacred. Sound is the thread that weaves together the spoken word, music, and silence into a tapestry of praise and reflection.

As you close this book, consider the next steps for your church. What can you do today to begin improving your audio ministry? Who can you collaborate with to build a stronger foundation? How can you inspire your congregation to value sound as an integral part of worship?

"Beyond the Sound Board" is more than a guide; it is a call to excellence and intentionality in church audio ministry. By valuing sound as a key part of worship, implementing practical changes, and envisioning a future of unified purpose, churches can transform the worship experience for generations to come.

Let us move forward with clarity and conviction, knowing that every adjustment made and every effort exerted in sound ministry has the potential to bring worshippers closer to God. The sound board may be a tool, but the ultimate goal is far greater: to create spaces where the divine is not just heard but profoundly felt. By striving for excellence in sound ministry, we echo the call to make a joyful noise unto the Lord, amplifying His glory for all to hear.

PRACTICAL CHECKLISTS TO ENHANCE AUDIO QUALITY

To ensure consistent audio quality in worship, practical and action-able steps are essential. Below are checklists organized into key areas of sound management that can help your church improve its audio output.

1. Pre-Service Audio Preparation Checklist

Task Category	Tasks
Room Setup	Verify all microphones and cables are properly connected.
	Ensure speakers are positioned correctly for optimal sound dispersion.
	Test acoustics and adjust temporary panels if needed.
Equipment Functionality	Power up all sound equipment (mixers, amplifiers, etc.).
	Confirm that batteries in wireless devices are fully charged or replaced.
	Check all devices for firmware updates.
Initial Sound Check	Perform a line check for every microphone and instrument.
	Set initial levels for vocals and instruments.
	Walk through the space to confirm even sound coverage.

2. Rehearsal Sound Check Checklist

Task Category	Tasks
Preparation	Confirm arrival of all musicians and vocalists.
	Distribute in-ear monitors or check monitor speakers for performers.
	Provide a sound system overview to new team members if applicable.
Testing	Run each instrument and vocal individually through the system.
	Adjust EQ settings to balance tones and remove feedback.
	Blend sound for full-band rehearsal, balancing all elements.
Feedback Loop	Solicit feedback from the worship leader and musicians.
	Fine-tune levels based on rehearsal dynamics.
	Make note of any persistent issues for troubleshooting.

3. Service Time Checklist

Task Category	Tasks
Pre-Service Final Checks	Confirm all channels are unmuted and at the correct levels.
	Ensure the room's ambient noise is managed (e.g., HVAC).
	Test communication tools (walkie-talkies, talkback mics) between sound team members.
Live Adjustments	Monitor levels consistently, especially during transitions (e.g., between songs, spoken word).
	Adjust for dynamics during worship to maintain balance.
	Resolve any live feedback or distortion issues promptly.
Post-Service Tasks	Power down all equipment safely.
	Note any technical issues encountered for follow-up.
	Archive the service recording, if applicable.

4. Troubleshooting Checklist

Issue Type	Tasks
Feedback Issues	Lower the gain on the affected microphone.
	Check speaker placement relative to microphones.
	Adjust EQ to reduce problematic frequencies (e.g., 2kHz-4kHz).
Distortion	Lower input levels on the mixer.
	Check for faulty cables or connections.
	Verify amplifier output matches speaker capacity.
Microphone Problems	Test backup microphones.
	Replace batteries in wireless systems.
	Confirm proper frequency settings for wireless devices.

5. Long-Term Audio Maintenance Checklist

Frequency	Tasks
Monthly Tasks	Inspect and clean all equipment.
	Test backup systems and spares.
	Evaluate sound coverage in the worship space.
Quarterly Tasks	Conduct a full inventory of audio equipment.
	Check for software or firmware updates for digital systems.
	Schedule team training sessions on system updates or new features.
Annual Tasks	Perform professional maintenance on key equipment.
	Update the audio ministry budget and plan for upgrades.
	Reassess worship needs and adjust audio strategies accordingly.

6. Designing Audio for Modern Worship Spaces Checklist

Design Aspect	Tasks
Acoustic Analysis	Conduct a professional acoustic assessment of the space.
	Identify and address sound reflection and absorption challenges.
	Plan for acoustic treatments such as panels or diffusers as needed.
Speaker Placement	Choose speakers suitable for the size and layout of the space.
	Position speakers to ensure even sound coverage for all seating areas.
	Minimize direct speaker alignment with walls to reduce reflections.
Microphone Selection	Use directional microphones for vocal clarity and feedback reduction.
	Consider boundary or hanging microphones for large group pickup.
	Ensure wireless microphones are on appropriate frequencies.
System Integration	Select a digital mixer with enough channels and functionality for growth.
	Integrate streaming or recording systems seamlessly with the sound setup.
	Include assistive listening systems for ADA compliance.
Future-Proofing	Design with scalability in mind for future upgrades.
	Incorporate flexible cabling and connection options.
	Plan for regular technology assessments and updates.

Scan the QR code below to get printable versions of the checklists.

7. Interviewing and Selecting Professional Audio Engineers Checklist

Selection Aspect	Tasks
Qualifications	Verify relevant certifications, training, and experience in live sound.
	Review previous work in churches or similar venues.
Technical Skills	Assess proficiency with digital and analog mixing consoles.
	Confirm familiarity with audio troubleshooting and repair techniques.
	Evaluate understanding of acoustics and sound system design.
Soft Skills	Ensure strong communication skills for working with worship teams.
	Look for adaptability and problem-solving skills under pressure.
References	Request and check references from previous clients or employers.
Trial Period	Offer a probationary period or trial run during a service.
Budget and Contract	Discuss salary expectations and contract terms upfront.
	Clearly outline job responsibilities and expectations.

By implementing these checklists, your audio team can maintain a high standard of sound quality, ensuring an engaging worship experience for all attendees. Printable versions of the checklists can accessed by scanning the QR code on page 128 or by visiting farmertechgroup.com/checklists.

APPENDICES

Appendix 1: Glossary of Church Audio Terms

Acoustics: The qualities of a space that affect how sound is heard within it.

Amplifier: A device that increases the power of an audio signal for playback.

Compression: An audio process that reduces the dynamic range of a sound signal to make it more consistent.

EQ (Equalization): Adjusting the balance between frequency components in an audio signal.

Feedback: The high-pitched noise caused when a microphone picks up sound from a speaker it is connected to.

Gain: The amount of amplification applied to an audio signal.

Mixer: A device that combines, processes, and adjusts multiple audio signals.

Phantom Power: Electrical power transmitted through microphone cables to operate condenser microphones.

Sound Check: A test conducted before a service to ensure all audio systems are functioning correctly.

Streaming: Broadcasting audio and video content live over the internet.

Appendix 2: Sample Worship Audio Plans

Standard Sunday Worship Service

- **Setup:** Ensure microphones, speakers, and monitors are connected and operational.

- **Rehearsal:** Conduct sound checks with worship leaders and musicians.

- **Sound Mix:** Balance vocals, instruments, and spoken word.

- **Streaming:** Test and monitor audio levels for online audiences.

Easter Service

- **Additional Needs:** Extra microphones for choir, rental equipment for larger spaces.

- **Preparation:** Schedule an extended rehearsal with a focus on transitions and special elements.

- **Execution:** Assign a dedicated team member to oversee audio for in-person and online audiences.

Christmas Production

- **Setup:** Stage microphones and wireless equipment for performers.

- **Coordination:** Collaborate with drama teams and musicians to synchronize cues.

- **Follow-Up:** Conduct post-event debrief to document lessons learned for next year.

Appendix 3: Recommended Resources and Tools

Books

- *The Sound Reinforcement Handbook* by Gary Davis and Ralph Jones

- *Great Church Sound:* by James Wasem

Websites

- Church Sound Media Resources (churchsoundcheck.com)

- Behind the Mixer (behindthemixer.com)

Training Programs

- AVIXA Certified Technology Specialist (CTS)

- Online courses from Church Technical Leaders (farmertech-group.com)

Tools

- **Digital Audio Workstation (DAW):** Pro Tools, Logic Pro X

- **Mixing Consoles:** Behringer X32, Yamaha TF Series

- **Microphones:** Shure SM58, Audio-Technica AT2020

- **Streaming Platforms:** vMix, OBS Studio, Wirecast

- **Encoders:** Black Magic Web Presenter, ATEM Switcher

By using these appendices, pastors, leaders, and media teams can access the practical tools and insights necessary for managing church audio systems effectively.

ABOUT THE AUTHOR

Dr. Timothy Brandon Farmer is a visionary leader, educator, and ordained minister dedicated to the transformative power of faith, education, and technology. Born and raised in St. Louis, Missouri, he has spent his life weaving together his passions for ministry, community service, and innovative solutions.

Dr. Farmer is a proud alumnus of Morehouse College, where he graduated as the top-ranking student in his education major. He further honed his expertise by earning a Master of Education in Curriculum and Instruction from Argosy University and a Master of Divinity in Church and Community Leadership from the Interdenominational Theological Center (Morehouse School of Religion). In 2022, he achieved a Doctor of Ministry degree from Candler School of Theology at Emory University, and he is currently pursuing a Ph.D. in Public Theology and Community Engagement at Hampton University.

In addition to his theological pursuits, Dr. Farmer is expanding his creative and technical skills through graduate studies at Berklee School of Music in audio production and at Savannah College of Art and Design (SCAD) in photography and graphic design. His multifaceted education underscores his commitment to blending artistry, technology, and faith.

Professionally, Dr. Farmer has served as an educator for Atlanta Public Schools and now works as an Educational Technology Specialist within the district's IT department. He is also the founder of Farmer Tech Group, a technology training and staffing firm that bridges the gap between innovation and practical ministry needs. As an ordained minister, Dr. Farmer serves on the pastoral staff of The Greater Piney Grove Baptist Church, where his dynamic leadership inspires spiritual and community growth.

Deeply rooted in his commitment to service, Dr. Farmer actively contributes to various organizations, including the Kappa Alpha Psi Fraternity, Inc., the Morehouse College National Alumni Association, the St. Paul Saturdays Male Leadership and Development Program, and the Martin Luther King, Jr. National Holiday Planning Committee.

Throughout his career, Dr. Farmer has received numerous accolades, including the INROADS Mark of Excellence Award, the Morehouse College Top Ranking Student in Education Award, and recognition as one of the Outstanding Young People of Atlanta. In 2024, was inducted into the Martin Luther King, Jr. Board of Preachers at Morehouse College, a testament to his profound impact on faith and community.

Through his work and writings, Dr. Farmer continues to inspire others to embrace the intersections of technology, faith, and community, creating a legacy of innovation and empowerment for generations to come.

www.ingramcontent.com/pod-product-compliance
Lightning Source LLC
Chambersburg PA
CBHW061650120626
46550CB00003B/892